SUSTAINABILITY

Sustainability, as a reference frame for dealing with the interconnection of environmental, economic and social issues on a global scale, is not only characterised by complex problems and long-term strategies but also by differences and disagreements with regard to its meanings and how they should be realised. Therefore, rather than seeking a single most appropriate definition of Sustainability, the main focus of this book is on how specific Sustainability problems are defined, by whom and in which contexts, what solutions are pursued to tackle them, and which effects they have in practice. This account of the social nature of Sustainability is intended to assist its readers to better understand the complexities, dynamism, and ambivalence of this concept as well as to find their own position in relation to it. For this purpose, the book traces the historical development of the larger discourse on Sustainability and investigates responses to three grand Sustainability challenges: climate change, energy, and agricultural food production. It suggests that promoting Sustainability requires continuous and active care and is inseparable from political debate about the normative foundations of society.

Thomas Pfister is Head of the EnergyCultures Research Group at Zeppelin University in Friedrichshafen, Germany.

Martin Schweighofer is a Researcher at the EnergyCultures Research Group at Zeppelin University in Friedrichshafen, Germany.

André Reichel is a Professor for Critical Management and Sustainable Development at Karlshochschule International University in Karlsruhe, Germany.

KEY IDEAS
SERIES EDITOR: PETER HAMILTON

Designed to compliment the successful *Key Sociologists*, this series covers the main concepts, issues, debates, and controversies in sociology and the social sciences. The series aims to provide authoritative essays on central topics of social science, such as community, power, work, sexuality, inequality, benefits and ideology, class, family, etc. Books adopt a strong 'individual' line, as critical essays rather than literature surveys, offering lively and original treatments of their subject matter. The books will be useful to students and teachers of sociology, political science, economics, psychology, philosophy, and geography.

Citizenship
Keith Faulks

Class
Stephen Edgell

Community – second edition
Gerard Delanty

Consumption
Robert Bocock

Globalization – second edition
Malcolm Waters

Lifestyle
David Chaney

Mass Media
Pierre Sorlin

Moral Panics
Kenneth Thompson

Old Age
John Vincent

Postmodernity
Barry Smart

Racism – second edition
Robert Miles and Malcolm Brown

Risk
Deborah Lupton

Social Capital – second edition
John Field

Transgression
Chris Jenks

The Virtual
Rob Shields

Culture – second edition
Chris Jenks

Human Rights
Anthony Woodiwiss

Childhood – second edition
Chris Jenks

Cosmopolitanism
Robert Fine

Nihilism
Bulent Diken

Transnationalism
Steven Vertovec

Sexuality – third edition
Jeffrey Weeks

Leisure
Tony Blackshaw

Experts
Nico Stehr and Reiner Grundmann

Happiness
Bent Greve

Risk – second edition
Deborah Lupton

Social Identity – fourth edition
Richard Jenkins

Knowledge
Marian Adolf and Nico Stehr

Renewable Energies
Matthias Gross and Rüdiger Mautz

Sustainability
Thomas Pfister, Martin Schweighofer and André Reichel

SUSTAINABILITY

Thomas Pfister, Martin Schweighofer and André Reichel

LONDON AND NEW YORK

First published 2016
by Routledge

2 Park Square, Milton Park, Abingdon, Oxfordshire OX14 4RN
711 Third Avenue, New York, NY 10017

Routledge is an imprint of the Taylor & Francis Group, an informa business

First issued in paperback 2018

British Library Cataloguing in Publication Data
A catalogue record for this book is available from the British Library

Library of Congress Cataloging in Publication Data
A catalog record for this book has been requested

ISBN: 978-0-415-71411-2 (hbk)
ISBN: 978-1-138-54635-6 (pbk)

Typeset in Bembo
by Taylor & Francis Books

CONTENTS

Abbreviations   *viii*

Acknowledgement   *x*

About the authors   *xii*

1   Introduction   1

2   Historical reflection: A brief genealogy of sustainable
development   10

3   Sustainability and climate change   26

4   Sustainability and energy systems   43

5   Sustainability and food systems   62

6   Sustainability as transformation and reflexivity   78

Bibliography   *94*

Index   *107*

ABBREVIATIONS

BSE	Bovine spongiform encephalopathy (also mad cow disease)
CO_2	Carbon dioxide
COP	Conferences of the Parties
CDM	Clean development mechanisms
CSR	Corporate social responsibility
DDT	Dichlorodiphenyltrichloroethane
EROI	Energy return on energy invested
EU	European Union
EU ETS	EU Emissions Trading System
GDP	Gross domestic product
GHG	Green house gas
GMO	Genetically modified organism
IAASTD	International Assessment of Agricultural Knowledge, Science and Technology for Development
IEA	International Energy Agency
IFOAM	International Federation of Organic Agriculture Movements
IPCC	Intergovernmental Panel on Climate Change
IUCN	International Union for the Conservation of Natural Resources

MDGs	Millennium Development Goals
NGO	Non-Governmental Organisation
OECD	Organisation for Economic Co-operation and Development
SDC	Sustainable Development Commission
SDGs	Sustainable Development Goals
UN	United Nations
UNCED	United Nations Conference on Environment and Development
UNEP	United Nations Environment Programme
UNFCCC	United Nations Framework Convention on Climate Change
USDA	United States Department of Agriculture
WCED	World Commission on Environment and Development
WCS	World Conservation Strategy
WHO	World Health Organization

ACKNOWLEDGEMENT

This book is the outcome of an on-going discussion its authors began when they first met in 2011 as researchers in the European Center for Sustainability (ECS) at Zeppelin University (ZU). The discussion continued even after one of us took up a professorship at another institution, Karlshochschule International University. We were coming from different directions in terms of disciplinary backgrounds, theoretical perspectives, and research interests. At some point, we had to find a shared outlook on Sustainability beyond our specific research interests. One particular way of addressing this concern was to reflect on the nature of Sustainability as an idea that is interpreted and enacted differently at different times and in different contexts. Against this background, the book is less the outcome of this discussion but rather the first attempt to tie its different strings together in a more systematic manner. Rather than closing the broader discussion amongst us, the book brought up several issues and problems which will make it worthwhile to continue the conversation.

Academic books are never written in a void. More people are involved in the publication process as well as the working environments of their authors. We want to name two persons in particular. We want to thank Manfred Moldaschl, director of the ECS, for caring for a working environment that is intellectually fruitful and built on good interpersonal relationships. In turn, Alyson Claffey at Routledge used a perfectly balanced mix of patience and insistence to keep us on track.

The ECS received financial support from Rolls Royce Power Systems (formerly Tognum AG). Thomas Pfister and Martin Schweighofer also gratefully acknowledge the financial support from the German Federal Ministry for Education and Research (grant number *01LN1312A*). André Reichel gratefully acknowledges the support of Karlshochschule International University for allowing enough time off from other academic obligations.

ABOUT THE AUTHORS

Thomas Pfister is Head of the EnergyCultures Research Group at Zeppelin University in Friedrichshafen, Germany. He holds a Master's degree in Political Science, Sociology, Modern History from Ludwig Maximilian's University Munich and a PhD in Political Science from Queen's University Belfast. His main research interests are in the relationship between science, society, and politics, particularly with regard to sustainable development, energy transitions, and European integration. In particular, he is interested in the epistemic dimension of governance in these contexts and its interaction with concepts and practices of citizenship.

Martin Schweighofer is a Researcher at the EnergyCultures Research Group at Zeppelin University in Friedrichshafen, Germany. He holds a Master's degree in Economics from the University of Vienna, has worked as a sustainability researcher and project manager for several years, and is now heading for his PhD in Sociology. His main research interests are in cultural transformations towards sufficiency and the respective dynamics of social practices, especially looking at the epistemic dimension. Currently he is doing research on the importance of social movements in the transition of energy cultures.

André Reichel is a Professor for Critical Management and Sustainable Development at Karlshochschule International University in Karlsruhe, Germany. He holds a Master's degree in Management and a doctoral degree in Economics and Social Sciences from the Universität Stuttgart. His main research interest is on degrowth and postgrowth with a special emphasis on microeconomic actors like companies and civil society organisations. More information about his work can be found at www.andrereichel.de.

1

INTRODUCTION

The idea of Sustainability has become a global norm that is adopted and pursued by a vast number of people and organisations worldwide. International organisations, national, regional, and local governments assumed this idea as a political objective and developed strategies and practices for its realisation. Businesses publish Sustainability reports and have specialist staff to monitor their ecological and social impact. They practice corporate social responsibility (CSR) and market specific products on the basis of their Sustainability. Moreover, many scientific disciplines have integrated the analysis of challenges to Sustainability and possible responses in their research agendas ranging from meteorology and oceanography to engineering, economics, anthropology, and science and technology studies. Finally, non-governmental organisations (NGOs) and social movements play a particularly important role in defining the meanings of Sustainability, key challenges, responses, as well as in scrutinising governments and businesses in terms of Sustainability. On the whole, the idea of Sustainability is meanwhile so embedded in the agendas and everyday practices of those agents that not many dare to claim openly that they would prefer to be unsustainable or do not care at all.

However, what exactly does it mean if, for example, a society is aiming for Sustainability? Does it mean that this society should focus on becoming 'carbon neutral' with regard to the operations of its economy and the everyday activities of its citizens? Or should it pursue more ambitious goals and seek to reduce its overall ecological footprint by not using more resources than the world's ecosystems can reproduce and by not causing more emissions and waste than they can absorb? Would such a development be possible by 'greening' the currently dominant modes of production and consumption through technological innovations and ethical consumption? Should our societies more generally seek technological solutions to the most pressing Sustainability challenges? How should we assess risky and uncertain technologies such as nuclear power, climate engineering, carbon capture and storage, or genetically modified organisms (GMOs)? Or is promoting Sustainability rather about transforming life styles, everyday practices, and culture? Would Sustainability even require a more fundamental re-orientation away from a globalised economy continuously seeking to maximise growth? Would sustainable development not also include a strong element of social justice? Moreover, would it also not imply that politicians should consider the impact of their decisions on future generations? And what about those living in the poorer parts of the world? Does the idea of Sustainability imply that societies redirect some of their wealth in order to increase the well-being of those people who are much poorer? Is it just basic common sense to be sustainable or are we living in times of much more pressing crisis and urgency? Finally, one important aspect must not be forgotten: the actual effects of the idea of Sustainability remain limited. Ecological pressures and social disparities persist and in many cases increase. Hence, one could also ask whether sustainable development is a useful concept at all or whether it is not more than a rhetorical frame people use strategically to make them look responsible or naively to feel good.

It seems that each of these questions has its relevance and that each question captures important aspects of Sustainability and how this concept is used in real world interactions. On the one hand, Sustainability is a concept that has entered the everyday life of politicians, bureaucrats, business managers and activists from the global to the local level. On the other hand, even a brief first look at the multiple contexts in which it is used, reveals a multiplicity of understandings and activities. Helping readers to make sense of this plurality of uses of Sustainability and to navigate more comfortably between them is a central aim of this book.

The book outlines Sustainability as a key idea that informs contemporary social practice and order. This idea is characterised by the complexity of the problems and goals it aims to capture. Moreover, this implies that Sustainability comes with different meanings and is related to different practices at different times, in different places, by different actors. To make matters worse, this implies that attempts to define challenges to Sustainability and to promote responses regularly run into tensions and conflicts with others – even if they are not generally opposed to the idea. In other words, Sustainability is an essentially contested concept (Gallie, 1956), which means that its complexity, variability, and interpretive flexibility will stay and continue to create tensions, contradictions, and conflict. Therefore, this book aims to make sense of these characteristics rather than attempting to prescribe one particular definition or set of related activities. At the same time, the book does not attempt to provide a panoptic account of all present and historical meanings used since this idea entered the global normative landscape. Instead, it operates at a slightly more abstract level offering an account on the social nature of Sustainability that explains its hybridity, dynamism, and ambivalence.

By now, it should be clear that this brief account is mainly written from a perspective of the social sciences. Hence, the book treats the idea of Sustainability as a social phenomenon and struggles to overcome related challenges and to promote transformations towards this goal as essentially social processes. Of course, most challenges to Sustainability are closely connected with the materiality of natural environments and human artefacts, for example, the atmosphere, oceans, soil, and plants, as well as photo-voltaic panels, electricity grids, and nuclear reactors. Moreover, the social science perspective fully acknowledges the importance of the natural sciences in this context. In order to gain thorough understanding of the most pressing problems of un-sustainability and to formulate strategies to counter them is in many cases impossible without the sophisticated expertise of natural scientists and engineers. However, even the most sophisticated models and most certain knowledge do not determine appropriate actions.[1] In contrast, a perspective from the social sciences emphasises exactly those (inter)actions that are necessary when people define complex and large-scale problems, when they formulate values and objectives they want to achieve, and when they search for ways how to realise them. Such situations always involve room for (differing) interpretations, different options, choices, norms, and interests as well as winners and losers. In short, the way – or transformation – towards greater Sustainability is always political (see also Scoones, Leach, & Newell, 2015).

More specifically, this book approaches the key idea of Sustainability with a focus on how it operates in practice. Therefore, rather than seeking a single most appropriate substantive definition of Sustainability, our interest is more on how specific Sustainability problems are defined, what solutions are pursued to tackle them, and which effects they have in practice. Sometimes they cooperate but often they find themselves in competition with other understandings and actors. Moreover, such attempts to become more sustainable will rarely be restricted to a single sphere, say politics or science, but mostly extend across different arenas at the same time.

In theoretical terms, this view on transformation projects and Sustainability in practice is grounded in the broader and more general discourses about practice theory. Theories of social practice have a long tradition and can be found in many nuances in the social sciences. The illustrious group of authors contributing to this body of literature includes, for example, Pierre Bourdieu, Anthony Giddens, Judith Butler, Harold Garfinkel, Karin Knorr-Cetina, Theodore Schatzki, Bruno Latour, and Michael Lynch. Their theories differ in emphasis and detail but all of them conceptualise practice as the basic unit of social analysis – in contrast to individuals or social structures. According to Andreas Reckwitz,

> A practice (…) is a routinised type of behaviour which consists of several elements, interconnected to one other: forms of bodily activities, forms of mental activities, things and their use, a background knowledge in the form of understanding, know-how, states of emotion and motivational knowledge.
>
> *(Reckwitz, 2002, p. 249)*

More recently, practice theories have been applied to Sustainability issues, but mainly to show the difficulties to transform arrangements of practices that produce Sustainability problems as side effects, for example, when the air conditioning and heating have become more or less invisible infrastructures required for the smooth performance of practices of clothing (mainly in suits) in office environments (Shove, 2003). On the other hand, social practice theories are not rigid or static. Especially, academics from science and technology studies have demonstrated the relevance of practices for the production of new knowledge and technologies (Knorr Cetina, 2001; Latour, 1987; Lynch, 1993). All of those more recent uptakes of practice with regard to Sustainability as well as science and technology also emphasise the embeddedness of practice within the material world. For example, farming practices developed differently in different places because they have to deal with very different environmental conditions such as weather, soil, and temperature. Moreover, in many cases, material and social aspects will overlap in farming practices, for example, in a large farm, which produces for the world market with large machineries and resource input.

Negotiations at a United Nations climate conference, decentralised energy production by local energy cooperatives, or the certification of fair trade products are all different practices, each consisting of specific elements, knowledge, meanings, and material elements (see Shove, Pantzar & Watson, 2012). On the one hand, human agents need to know what situation they are in and how to act accordingly – which practices to enact. On the other hand, this requires that others are able to recognise a specific practice even if its meaning is not explicitly stated. This is possible, when practices achieve a degree of regularity or, in other words, constitute social order. Moreover, practices can be disrupted by external events, changes in the material world, or by other practices intentionally or unintentionally intervening in a particular order. These could, for example, be practices of innovation, critique, and resistance. All of them are relevant with regard to Sustainability. Building an order that is more sustainable requires establishing transforming and establishing extensive regimes of different practices that implicitly or explicitly address particular Sustainability problems. Moreover, given the political nature of such processes, critique, contestation, compromise, and other innovations remain essential elements throughout (otherwise they would break down).

How does this theoretical basis influence the understanding of Sustainability as an idea? Sustainability needs to be thought of as embedded in different social practices, for example, practices of writing academic papers, newspaper articles or blog posts, giving speeches, or holding debates is the material one has to investigate what specific actors mean when referring to Sustainability. In the words of Ludwig Wittgenstein, 'the meaning of a word is its use in language' (1953 no. 43) – how it is used in practice. Hence, if we are speaking about meanings and practices of Sustainability, we still assume that meanings are part of and embedded in practice. In addition to these discursive practices of defining, regulating, elaborating, and contesting the notion of Sustainability, there are practices people pursue who want to be more sustainable without making it explicit. For example, some people use public transport rather than cars, some farmers adopt practices of organic farming, and some scientists gather satellite data to feed computer simulations and to determine particularly critical thresholds after which the negative effects of global warming would increase significantly. This points to the more general issue of all those mundane practices from cooking, to farming, from vacationing to global logistics, which could be more or less sustainable. They pose serious and often very difficult questions with regard to how certain Sustainability challenges could be overcome and how unsustainable societies could be transformed.

In order to make sense of the multiple and contested uses of Sustainability and related practices, we trace, for example, phases and contexts when certain meanings and implicit practices have stabilised. In this manner, the next chapter identifies several major phases of the global debate on Sustainability, each related to particular concerns, objectives, problem analysis, and agents. Moreover, Chapters 3–5 focus on more specific challenges to Sustainability and identify several stabilised constellations of more specific practices and meanings as well as specific agents promoting them. Such more specific contexts could be described as specific 'projects' carried by certain networks of agents, tackling different challenges to Sustainability, and following different meanings and aspects of this idea. They can develop around a common language, institutions, shared procedures, or specific technologies and infrastructures.

Before this notion of Sustainability in practice is further explored, the remainder of this chapter introduces some key terms and explicates the focus, structure, and omissions of this rather short treatise. To begin with a working definition, Sustainability can be understood in a narrow physical sense, for example, as an equilibrium between all material flows that go into the economy (i.e. resources) and the material output in terms of waste and emissions. This is, for example, the view on the earth as a closed ecological system that was prominent in the report *Limits to Growth* or Herman Daly's physical conceptualisation of a steady-state economy embedded in a limited biosphere, where throughput and material output of the economy are in balance with input of non-renewable resources (Daly, 1977).

However, in 1987, when the World Commission on Environment and Development (WCED) presented its famous report *Our Common Future* (often cited as the 'Brundtland Report' after the Commission's chair Gro Harlem Brundtland) and introduced the concept of *sustainable development* to the global policy arena, it included much more than physical analysis and material flows. Instead, it defined sustainable development in terms of human needs and equity creating one of the most influential definitions to the present day: 'Sustainable development is development that meets the needs of the present without compromising the ability of future generations to meet their own needs' (WCED, 1987 part I, chapter 2, no. 1).

As we will discuss in the next chapter, the history of Sustainability is to a lesser extent about the emergence of a new idea but rather about new connections and realignments of several ideas that have been established for a very long time. Ecological concerns about environmental impacts of human activities on the Earth's ecosystems are the most recent and central source of the Sustainability discourse. Concerns about persisting global inequalities constitute another core issue.

Despite the seriousness of these disparities between global North and South and despite the importance of related concerns, this book will mainly use the more general term *Sustainability* (not generally including the development aspect). In part, this reflects the main focus of this book on Sustainability challenges mainly caused by western/northern states. For the greater part, however, this use reflects the various degrees and shifts in meanings, emphasis, and composition that the key idea of Sustainability has undergone since its introduction to global politics by the WCED. In fact, in many instances global equity is less prominent than environmental aspects of Sustainability. At the same time, issues of global action, distribution, and equity feature prominently throughout the whole study. Finally, in order to distinguish the main topic of this book from more colloquial uses of the word such as in a 'sustainable budget' or a 'sustainable competitive advantage', we follow Andy Stirling (2009) by using Sustainability with a capital 'S' as a more specific key idea of human thought.

Overall, enquiring about the meaning of Sustainability as a contemporary key idea requires attention to the various meanings-in-use and practices that are mobilised by different agents in order to realise it in specific contexts. Moreover, this contextualised and practical nature of Sustainability requires that related challenges and struggles are broken down in more detailed enquiries but also re-assembled into larger pictures of general crosscutting themes and complex problems. The following chapter rather moves in the latter direction, describing the history of Sustainability in terms of larger patterns that have remained dominant at particular times. Then, Chapters 3, 4 and 5 present more detailed investigations of specific themes, while the final chapter returns to reflecting on the key idea of Sustainability as a whole.

As mentioned, Chapter 2 takes a historical look at the development of the larger discourse on Sustainability. It inquires how particular meanings and elements of Sustainability became dominant at certain times and distinguishes three main phases. This historical narrative is far from linear but characterised by significant junctures, a consistent plurality of different outlooks, and essential contestedness even within phases of longer stability.

The next three chapters look at three grand Sustainability challenges: climate change, energy, and agricultural food production. The chapters largely follow the same structure. They analyse the main problems in these contexts from a Sustainability perspective, illustrate how certain projects respond to these challenges, mobilising specific ideas of Sustainability and putting them into practice – at times in very different ways.

The final chapter takes a more general view on Sustainability as a key idea again drawing conclusions about its composite, fragile, and contested nature. Moreover, it reflects on the potential gains and limits of Sustainability as an important idea in the contemporary world. It argues that the most important role of Sustainability is not as a (physically defined) equilibrium that could be achieved or an endpoint to be achieved. Rather, Sustainability as an idea should be seen as a much more dynamic and open-ended social process that provides an important source of reflexivity for today's industrialised, energy intensive, consumerist societies and a world where access to well-being is distributed very unequally.

However, the links between climate change, energy, food, and Sustainability are so manifold and complex that these chapters can only scratch at the surface of those issues added and have to leave many questions open and many aspects unmentioned. Moreover, the list of possible further challenges to Sustainability is without end. For example, the book could also have discussed questions of biodiversity, water, waste, mobility, deforestation, urbanisation, finance, health, or private consumption to gain equally relevant insights. Moreover, we could add another equally long list of agents, institutions, policies, strategies and instruments addressing one or several of these challenges. Yet, the book is meant to provide a rather short conceptual introduction not an exhaustive empirical study of all aspects of Sustainability. On the one hand, it should help to better understand its specific character as an intellectual framework consisting of diverse and contested elements. On the other hand, this framework is closely connected with material aspects of resource use, outputs such as waste and emissions, and the world's ecosystems. It is the world where challenges are detected but it is the idea of Sustainability that makes them visible, urgent, and provides orientation in response to them.

Note

1 It is impossible to elaborate on this point in the confined space of this book. Instead, we have to hint at the extensive body of literature, primarily in science and technology studies, contributing to this discussion (for example, Felt et al., 2013; Forsyth, 2003; Hulme, 2014; Jasanoff, 2012; Nelkin, 1979; O'Riordan, 2004; Pielke, 2007; Sarewitz, 2004; Stirling, 2009; Wynne, 2010)

2

HISTORICAL REFLECTION

A brief genealogy of sustainable development

Introduction

If one is interested in the workings of Sustainability as an idea, one could start the investigation by comparing it to other key ideas that structure actions, expectations, and judgements in many communities, states, and organisations around the globe such as democracy, justice, and human rights. Such a comparative view is a helpful first step towards understanding the different meanings of Sustainability that were hinted at in the introduction and that are to be encountered in the following chapters.

Therefore, the discussion of this chapter begins by briefly looking at democracy. Although most people might have an intuitive sense what democracy means, it comes in many shapes and varieties once we are looking beyond a single democratic jurisdiction. The history of political philosophy offers quite a wide range of different democratic theories, which can be labelled, for example, as liberal, representative, republican, direct, discursive, and radical democracy. They all might share fairly general ideas of political self-determination, in particular through voting as the most important exercise of political voice. However, on closer inspection, the various democratic theories can differ significantly. They can be in tension or in contradiction to each other. Similarly, if we look at the institutions of democracy, we quickly realise that it is hardly possible to find two governments or parliaments that are identical (the whole academic discipline of comparative political science lives on this diversity). Moreover, when people protest against situations where they feel powerless, excluded, or oppressed, they often voice their criticism in terms of violations of democratic rights or basic principles and make claims for more democracy. While political protests can be seen as a further aspect of democratic practice, they often also feed back into democratic thought and institutions when they are redrawn in response to these protests. More recent deliberative, feminist or agonistic democratic theories are closely connected to the activities and values of social movements and the contestation of traditional political elites and majoritarian political institutions. In short, even (perhaps, especially) key ideas such as democracy acquire different meanings in different contexts; they are inscribed in very different institutions and practices, and they are essentially contested.

As will be shown in the course of this and the following chapters, Sustainability shares these characteristics. In addition, however, it has a very distinct feature that makes diverse meanings, practices, and contestations even more likely: its *composite* nature. Sustainability might be a relative newcomer to the normative architecture of the modern globalised world. However, the contents of this new idea were not entirely unknown. Rather, they existed as individual ideas before and there also is a history of thinking how they interact with each other. What was novel when Sustainability was invented was the way these ideas were recombined. Moreover, the novelty of this recombination was only in part an intellectual development, it was also a significant political reorientation (Dresner, 2012).

Understanding Sustainability as an idea is also helpful because it implies that this concept is neither an objective description of the world nor necessary consequence of the human impact on it. Instead, it is always necessary for someone for certain reasons. Yet, some ideas are so widely accepted that they are perceived to be necessary or 'true'. And most importantly, this taken-for-grantedness cannot be found in the idea itself but in its history as a social and cultural phenomenon, for example, when people use a concept to make sense of the world, to direct, to criticise, and to legitimate action. Therefore, this chapter does not investigate how sustainability was 'discovered' in history but rather which different versions came to be seen as more accurate than others and how this happened. Since Sustainability is a composite idea, it also traces its different developments before the compromises and innovations that led to this composition. Moreover, this historical narrative is not about a linear development but about phases of stabilisation as well as about ruptures and phases of destabilisation. It identifies three stages in which particular constellations of meanings and practices related to Sustainability were dominant. This understanding in terms of main phases does not deny that Sustainability is always open to different interpretations. At the same time, the more general framing of Sustainability as the need to balance economic, social, and environmental objectives and actions remains its more general framing.

This historical account operates at a middle range looking for stability of an idea and related practices that is at the same time specific but plural and contested on closer inspection, as well as relatively coherent but vague at the most general level. This focus on patterns of stability and processes of stabilisation also opens up a view on power and the political nature of Sustainability and ideas more generally. This implies that attention should be paid not only to the historical development of the contents of the idea of Sustainability but equally to the social and political conditions and forces promoting those contents that attract the strongest support, are most likely to be seen as 'normal', 'objective' or even 'true'.

From a methodological point of view, investigating how ideas are historically made powerful and perceived as true is also described as *genealogy*.[1] A genealogical perspective does not aim at uncovering a single most truthful interpretation of historical events but is interested in the turning points and discontinuities that lead to dominant constellations and also in the alternative options and perspectives that struggle along at the margins as minority opinions, political opposition, or sub-cultures. Since a meticulously detailed intellectual history of Sustainability would go against the concise character of this book, we restrict ourselves to roughly sketching three major phases when different versions of Sustainability were paradigmatic (following Kuhn, 1962).

Human life, and human well-being essentially depend on the use of natural resources. In this context, the relationship between human economic activity, the environment, and human well-being has been the topic of human thought for a very long time. For example, Du Pisani (2006) mentions Roman authors like Lucius Junius Moderatus Columella (4 AD to 70 AD) and Marcus Terentius Varro (116 BC to 27 BC) who investigated – primarily with regard to farming – how negative environmental effects of resource exploitation and economic activity could be mitigated through less exploitative farming practices. This points to a relationship between humans and the environment characterised by a need to *care* for the latter. In contrast, Thomas Robert Malthus (1766–1834) made much more rigid assumptions about the limited nature of natural resources leading to his very pessimistic assessment of population growth. In addition to sharing Malthus's fear of overpopulation, John Stuart Mill (1806–1873) also argued that a *stationary state* should be preferred over continuous growth because the latter would eventually undermine the natural basis for good living conditions. However, also the difficulties implied by this need to care for the environment and in particular for natural resources were already discussed in the nineteenth century. A particularly important idea in this respect is William Stanley Jevon's (1865) work on 'rebound effects', which describes his finding that the invention of more efficient steam engines by James Watt nevertheless led to much higher overall consumption of coal (see also the discussion about peak oil in Chapter 4).

Given the importance of wood as the central natural resource since prehistoric times, forest management is an area where much thought has been devoted to avoiding overexploitation and ensuring sustainable yields. Forestry also was the intellectual home of several writers who are regularly described as ancestors of the contemporary Sustainability discourse, such as John Evelyn in England or Hans Carl von Carlowitz in the German Electorate of Saxony. The latter coined the principle of sustainable forest management to cut only as much timber as the forest can reproduce in his treatise 'Sylvicultura Oeconomica' in 1713. In fact, he used the German expression for 'sustainable' to characterise this way of forest management. More generally, the notions of sustainability were used in the German, French, or Dutch language for centuries, while the word entered the English language only in the second half of the twentieth century (Du Pisani, 2006). These notes on the 'prehistory' of Sustainability do not imply a linear development towards the contemporary debate about Sustainability. However, they illustrate that the idea is not entirely novel, since the language and further related practices have existed before. At the same time, these historical fore-runners do not provide the most important source although regularly mentioned in contemporary debates. This part of the history of ideas is only one aspect within a more diverse field of meanings, practices, and historical reference points. For example, the German historian Joachim Radkau (2011) states that the green movements in Germany and the USA did not feel a particular connection to predecessors of this kind. In particular, German greens were very critical of earlier traditional, romantic, and, at worst, national socialist strands of environmentalist thought. The diversity of possible sources, legitimations, and practices was too broad and too ambivalent to provide a satisfactory narrative.

Imagining the world as a finite ecosystem and livelihood in need of care

While it is important to be aware of these historical contexts, we begin the main part of our historical narrative in the 1960s when a new para-digmatic view on the environment took hold and spread across a global scale. It is characterised by a growing awareness of the world as a fragile ecosystem, the limited nature of its resources, and the potentially dramatic consequences of its overexploitation by human activity. Moreover, this awareness was translated into first institutionalisations, in particular, non-governmental organisations (NGOs) and a United Nations (UN) programme. Importantly, while the terms Sustainability or sustainable development were not yet prominent, this phase established the envir-onment as a sphere that was closely interlinked with human economic activity and well-being and, therefore, required at least similar political attention.

A first symptomatic event of this phase was Rachel Carson's *Silent Spring* (1962). Her book focused primarily on the catastrophic effects of the widespread use of pesticides, in particular, dichlorodiphenyltri-chloroethane (DDT). Although its main focus was on the negative effects on animals it also drew much attention to the harmful consequences for humans. Based on a fine balance between scientific analysis and powerful literary imagination *Silent Spring*, thereby, not only became a best-selling ecological classic but also helped to establish the environment as a public issue. Carson's book was seconded by Garret Hardin's article in Science 'The tragedy of the commons' (Hardin, 1968) or the 'Blueprint for survival' in the *Ecologist* magazine (1972), which also contained the first ever newsletter by Friends of the Earth (founded in 1969).

At the centre of this first phase, however, we also place the first report to the Club of Rome *Limits to Growth* (Meadows et al., 1972). In contrast to Carson's balance between scientific detail and literary skill, the report captured the imagination mainly by the sheer scale of the underlying scientific endeavour. It was based on a computer simulation to explore the interactions between (population and material rather than economic) growth and the finite resources available on the planet. On this basis, the authors of the report drew up several scenarios, most of them based on overshooting ecological limits and producing devastating consequences for global health, prosperity, and the environment ('collapse'). However, the simulation also demonstrated that a 'global equilibrium' (Meadows et al., 1972, p. 24) was possible.

The report was heavily criticised from the beginning, with criticism ranging from refutations of the underlying methodology to general condemnations on ideological grounds. At the same time, it sold over 12 million copies, was translated in 37 languages, and followed by regular updates keeping alive the hotly contested debate. With regard to the history of the idea of Sustainability, the reception of *Limits to Growth* is especially interesting since it allows for studying the main assumptions and concerns characterising this first phase as well as the conflicts and struggles they provoked. A key assumption, in this context, was that resources and absorption capacities of the world are limited and that humanity was on its way to overshooting them. The resulting ecological crisis would, therefore, require an urgent and more or less radical break with dominant patterns. Its strict scientific framing and its basis in the most recent developments in the very young discipline of computer science at the Massachusetts Institute of Technology significantly contributed to this success, including academic debates, for example, about data and methods. Moreover, the oil crises of the 1970s seemed to provide instant empirical proof for these claims. Finally, based on the simulation of the world as a single dynamic system, the report drew on and contributed to an imagination of the world as finite, vulnerable, and enclosed system – the 'spaceship earth' (see, for example, Boulding,

1966) that was so powerfully represented by the first images of the earth from space (Jasanoff, 2001).

Although the term sustainable was not used to describe the inter-dependencies between environment, economy, and human well-being, the phase around *The Limits to Growth* was very relevant for the further historical development of this idea. On the one hand, Sustainability is regularly equated with environmental issues despite attempts to pro-mote a more comprehensive notion of sustainable development centred on equity (see next section). On the other hand, the discussions about *Limits to Growth* already illustrate the characteristic contestedness of Sus-tainability. This involves the tension between calls for greater restraint versus expectations of technological solutions to growing ecological footprints. Moreover, debates and conflicts relating to Sustainability appear in quite different spheres at the same time. The debates surrounding *Limits to Growth* and also *Silent Spring* were about values and morals, about scientific evidence and method as well as about political and economic interests.

Also in this early phase organised agents and institutionalised spaces started to emerge. Organisations such as Friends of the Earth (US: 1969, international: 1970) or Greenpeace (1972) originated in the broader social movements concerned with environmentalism, animal welfare, and anti-nuclear protest. Moreover, the first Green party was founded in England and Wales in 1973.

In addition to organisations emerging from social movements, expert knowledge on environmental issues became an increasingly important theme in academic as well as in policy circles. For example, the National Academy of Sciences in the USA published the first report on the impact of CO_2 on the global climate (Charney et al., 1979). Moreover, in 1980 the 'Global 2000 Report to the President Jimmy Carter' (Barney, 1980) translated the analyses and concerns of that era into concrete policy recommendations. However, these attempts were abruptly stalled after the election of Ronald Reagan and when com-peting ideas about market deregulation gained dominance over those environmental concerns.

At the international stage, however, a more sustainable regime emerged to provide the basis for the 'birth' of Sustainability and its global spread. The United Nations Conference on the Human Environment held in

1972 in Stockholm led to the foundation of the United Nations Environment Programme (UNEP). In the process, the UN became a central arena for global environmental policy and its integration into the more comprehensive idea of sustainable development.

The global rise of sustainability: from environmental angst to equity to economics

The second phase in our historical account of Sustainability is particularly characterised by three aspects: first, the introduction of the specific term 'sustainable development'; second, its quick adoption in international law and politics and its diffusion to national, regional, and local levels of government, the business world, and civil society (where it was not so new); and, finally, the strong and dominant framing of Sustainability in terms of global equity and poverty reduction. In the process, the terms Sustainability and sustainable development are used synonymously only for this part of the narrative.

The first and defining instance of this phase is the publication of the report *Our Common Future* by the UN's World Commission on Environment and Development (WCED), which brought the notion of sustainable development to the attention of the wider world. It did not invent it, though. Simon Dresner identifies the origin of the term in 1974 at a development-related conference by the World Council of Churches, which called for a 'sustainable society' that should be equitable, democratic, and have no negative impact on the environment (Dresner, 2012, p. 32). In 1977, the International Federation of Organic Agriculture Movements (IFOAM) held its first conference using the title 'towards a sustainable agriculture' (see also Chapter 5). In 1980, the International Union for the Conservation of Natural Resources (IUCN) launched its new World Conservation Strategy (WCS) with the subtitle 'Living Resource Conservation for Sustainable Development' (IUCN, 1980). However, it was the publication of *Our Common Future* – also named the Brundtland Report after the commission's chair, former Norwegian prime minister Gro Harlem Brundtland – that established this concept at the stage of world politics and proved to be so decisive for the specific meanings and practices of Sustainability during this phase. The report departed from the more dramatic shrink or perish-imaginary that was dominant in the previous phase and promoted sustainable development as a more moderate alternative that would nevertheless require the 'progressive transformation of economy and society' (WCED, 1987 part I, chapter 2, no. 3). Its definition of sustainable development as 'development that meets the needs of the present without compromising the ability of future generations to meet their own needs' (WCED, 1987 part I, chapter 2, no. 1) is still one of the best known definitions. However, in order to fully understand the meaning of Sustainability that we hold to be characteristic for that period, it is necessary to look at the rest of this definition that is usually not quoted. Accordingly, the second sentence reads:

It [sustainable development] contains within it two concepts:

- the concept of 'needs', in particular the essential needs of the world's poor, to which overriding priority should be given; and
- the idea of limitations imposed by the state of technology and social organization on the environment's ability to meet present and future needs.

(WCED, 1987 part I, chapter 2, no. 1)

In contrast to the serious concerns about the global environment, the focus in this high phase of sustainable development shifted towards the massive differences between the global North with its long history of overshooting global ecological limits and the South where most of the world's poor whose essential needs should be met are based. The meaning of growth is particularly affected by this shift. Instead of being imagined as an environmental risk, it came to be seen as an instrument to help those in greatest need. In contrast, the North was seen as required to drastically reconsider its growth (or to decouple its economic growth from material growth if at all possible) since most essential needs are already met. Yet, the meaning of growth as a goal in itself never disappeared and resurfaced in the more recent stage of the global Sustainability debate (see next section). One important aspect that aided the coming together of environmental protection and development into a single concept was the widespread disillusion with traditional notions of modernisation and progress among environmentalists as well as among advocates of the postcolonial South.

The Brundtland Report was crucial for outlining and stabilising the concept of sustainable development as well as for putting it in the context of human rights, equity, and poverty alleviation. In 1992, the United Nations Conference on Environment and Development (UNCED) in Rio de Janeiro, the so called *Earth Summit* established sustainable development as a central political norm and the UN as the main normative entrepreneur behind it. Regarding the outcomes of the summit, Principle One of the Rio Declaration on Environment and Development emphasises that 'human beings are at the center of concern for sustainable development' (United Nations, 1992a). The Agenda 21 outlined a plan for action based on a bottom-up and participatory approach particularly admitting NGOs and local actors to the political field. Finally, the United Nations Framework Convention on Climate Change (UNFCCC) was a first step towards the agreement of the Kyoto Protocol in 1997. In the process, climate change replaced the more comprehensive ecological consequences of material growth as the key environmental theme.

However, although the Brundtland Report and the Earth Summit met a crucial window of opportunity at the end of the cold war, their political consequences fell far behind the initial hopes and demonstrated the contestedness of sustainable development and the difficulties to put it into practice (see especially Dresner, 2012). Most outcomes were not binding but voluntary and met much resistance from the developing world, which opposed environmental regulation as obstacle to development, as well as from industrialised states in the North (in particular the USA).

This phase of conceptual definition and clarification also saw the publication of John Elkington's *Cannibals with Forks* (Elkington, 1997), which made the idea of the *triple bottom line* known to a global audience. The triple bottom line is most often a graphic representation illustrating sustainable development as the intersection of three circles relating to the economy, the environment, and social issues. This imagery, developed ten years earlier by Edwards Barbier (1987), did not only support the take-up of Sustainability beyond the political world. The language of bottom lines helps us to understand how the idea became operationalised and translated into practice. In particular, this task involved establishing and monitoring whether an activity would be sustainable or how much it would reduce unsustainability. This also involves the proliferation of numerous indicators, metrics and standards to account for Sustainability. In fact, accounting for Sustainability has become a central aspect of the repertoire of practices related to Sustainability – ranging from measuring CO_2 emissions, to monitoring biodiversity, to accounting for the social and environmental consequences of business activities, to assessing individual ecological footprints. On the one hand, Sustainability became operationalised and measurable in terms of a constantly growing range of indicators; this was crucial for its incorporation into policy and business practices. On the other hand, this rather technical (if not technocratic) development had the effect that this idea lost some of its power as a critique of the global political economy and as a normative imaginary of possible alternative worlds worth striving for.

While the concept of Sustainability-as-sustainable-development spread rapidly around the globe, this did not involve its realisation exactly as imagined by the Brundtland Commission. The report and the ensuing Rio Earth Summit introduced this idea in a strong framing of global equity and human well-being. However, the global conceptual debates and policy processes that followed could not uphold the dominance of this frame. Instead, the notion of Sustainability that entered the spheres of politics, business, and science at that time was more contested and open to differing interpretations and practices – many of them were geared towards measuring and accounting rather than putting in place the kind of radical transformations envisaged by the Brundtland Report. In the process, the initially strong focus on global poverty and the needs of the poorest became much weaker.

To conclude this section, the second stage of our genealogical narrative that is characterised by *Our Common Future* and the Rio Earth Summit can be seen as a phase of transition when the idea of Sustainability gained traction at a global scale, when it spread into national political debates as well as the worlds of science, business, and civil society, and became integrated in various institutions and triggered the emergence of new practices. However, despite the influential definition and the important underlying compromise, the equity centred notion of sustainable development could not keep up its dominance. Perhaps, this was an unintended consequence of its rapid diffusion; perhaps this meaning never had a chance because support was too fragile. In the process, the different aspects and dimensions of Sustainability have become assembled quite differently in different contexts.

The contemporary fragmentation of Sustainability

The third phase in our genealogy of Sustainability is a more hybrid situation. On the one hand, it is mainly characterised by a strong economic framing making the business case for Sustainability and promoting green growth. On the other hand, the dominant meanings and practices from the previous phases did not disappear completely. In the following, we will outline three ideal-typical constellations, which overlap and are contested in practice.

A first aspect concerns the growing awareness and perception of climate change as the most important environmental problem. The foundation of the Intergovernmental Panel on Climate Change (IPCC) by UNEP and the World Meteorological Organization in 1988, the signature of the UNFCCC at Rio 1992, and the signing of the connected Kyoto Protocol (1997) provided the institutional basis for this central role. At the same time, the immense difficulties in negotiating a successive agreement to Kyoto demonstrate that climate change is not only a major concern but also a major battleground in struggles for greater Sustainability. In particular, conflicts and struggles in this area are related to the contested interpretations of scientific evidence about climate change as well as to adequate political consequences. In the context of climate change and in part fostered by the hybrid role of the IPCC between science and politics, Sustainability is primarily framed as a complex scientific problem rather than a socio-political one of equity and well-being. In the process, political questions about vulnerability and costs, responsibility, and justice have often been overlaid with 'proxy debates' (Beck, 2011, p. 303) about scientific evidence for anthropogenic climate change (see also, Hulme, 2014; Wynne, 2010). Even concerning questions that are less complex than the human impact on global warming, the scientific framing implies that Sustainability is treated as a complex problem mainly addressed by experts but far away from life-worlds and everyday experiences. Climate change and Sustainability are articulated in terms of complex models, indicators, and data investigated and designed by scientists and highly skilled professionals. Given the importance of climate change in the context of Sustainability, we devote the next chapter entirely to this issue.

A second aspect concerns the adoption of Sustainability in mainstream practices of policy and business management. In this context, practices and meanings of Sustainability often draw on technical practices of measuring and accounting as mentioned above. Moreover, they are particularly influenced by the search for making the business case for Sustainability. A first landmark event in this regard motivated by the central concern about climate change was the *Stern Review on the Economics of Climate Change* (Stern, 2007). The review was commissioned by the UK Treasury and recommended massive and immediate action against climate change because the costs of inaction were estimated to be much higher. In the process, Sustainability was translated into yet another frame. In contrast to previous framings as systemic equilibrium or transformative project to achieve global equity, it was now reframed in terms or ecological modernisation and a green economy. This new meaning was based on the argument that Sustainability was not only necessary due to environmental and social crises but also reasonable from an economic point of view. For example, Thomas L. Friedman suggested a *Green New Deal* to tackle climate change and to achieve Sustainability roughly at the same time:

> we need more of everything: solar, wind, hydro, ethanol, biodiesel, clean coal and nuclear power – and conservation. It takes a Green New Deal because to nurture all of these technologies to a point that they really scale would be a huge industrial project.
>
> *(Friedman, 2007)*

This understanding of Sustainability spread quickly. For example, UNEP hopefully claimed that a new 'Green Deal generating businesses in renewable energies; clean tech ventures, sustainable agriculture, conservation, and the intelligent management of the planet's ecosystems and nature-based infrastructures is already under way' (UNEP, 2008). Even more important, these ideas proved to be particularly attractive when the financial crisis hit in 2008. For example, a research report commissioned by UNEP argued that:

> such a strategy is not just about creating a greener world economy. Ensuring the correct mix of global economic policies, investments and incentives can achieve the more immediate goals of stimulating economic growth, creating jobs and reducing the vulnerability of the poor and the long-term aim of sustaining that recovery.
>
> *(Barbier, 2009, p. 9; see also, UNEP, 2009)*

In the process, the Green New Deal became rebranded as *green economy*, much less closely associated with Keynesian ideas and the historical example of the US New Deal. From that perspective, Sustainability was imagined as key ingredient to remedy the global economic malaise. In other words, achieving greater Sustainability is not about fundamentally transforming the economy but about getting the economy right. Rather than being an alternative to the global economy the green economy became a means to succeed in the latter – a business model as well as a macroeconomic instrument to get the economy out of crisis and towards green growth. This imaginary is, for example, very visible in the EU's Europe 2020 strategy for economic growth, which shall be 'smart, sustainable, and inclusive'.[2] In this particular policy context, the main emphasis is on innovation rather than redistribution or environmental protection, and society as a whole is imagined as an economy rather than as a political order (Flear & Pfister, 2015).

The context of a green economy reconstructs issues of equity, global solidarity, and poverty reduction – core concerns in the phase characterised by the Brundtland Report and the Rio Earth Summit – by subsuming them under inclusive growth. This notion emphasises overcoming social and developmental issues by relying on market-driven economic expansion and its trickling down (World Bank, 2012). As mentioned, the framing of Sustainability as equitable sustainable development could never keep up its dominance over the practices and meanings of Sustainability at large – especially as they spread into very different spheres. Now, it is one aspect among several others and often of lower priority. At the level of the UN, it reappeared at the large international conferences on sustainable development in Johannesburg (Rio +10) ten years after the 1992 Earth Summit and again at the Rio +20 summit in 2012 in Brazil. Moreover, it is probably most visible within the Millennium Development Goals (MDGs) that were adopted in 2000. However, even the MDGs do not take up the equity-based notion of sustainable development originally put forward in *Our Common Future*. Rather, 'ensuring environmental sustainability' was adopted as seventh of eight development goals, which implies that the relationship between the overarching concept and its elements was turned on its head between 1987 and 2000. Fighting global poverty was the key priority of sustainable development in the Brundtland Report. In contrast, the MDGs are framed as fighting poverty and ensuring well-being while Sustainability is framed as a subordinate – and more narrow environmental – element. This does not mean that Sustainability does not play a role in this context. However, the combination of environmental, economic, and social issues under the heading of global equity was a major political achievement in the previous stage and was translated into a global political process legitimised by this conceptual imagination. With the MDGs, this close coupling was loosened again and the fight against poverty shifted to a more separate policy-process. Importantly, if someone wanted to point out potential problems within this development, one had to point to the decoupling and the move of equity, health, poverty reduction, and similarly fundamental aspects of human well-being to a single policy process and its exclusion from the policy mainstream dominated by notions of a green economy. Chapter 5 discusses further examples of this kind pointing to promoters of organic farming, fair trade, and food sovereignty who rediscovered a very tight connection between environment and equity.

Moreover, the weakening of the comprehensive view on sustainable development was partially due to symbolism motivated (and in parts made possible) by the calendar since the MDGs were adopted at the UN Millennium Summit in 2000. They were formulated for a period of 15 years. Hence, the UN convened a Sustainable Development Summit in September 2015 to adopt a new set of objectives, under the title *Sustainable Development Goals* (SDGs). The new list contains 17 SDGs, which are broken down into 169 more specific and measurable targets. While it is too early to speculate how this new agenda will be taken up in practice, it is important to note that the SDGs continue to constitute a space where the main focus is on global equity, human needs, and human well-being. Those goals explicitly relating to environmental concerns are mentioned further down the list. At the same time, the environmental aspects are integrated across all goals. In short, it remains to be seen whether and how this new and comprehensive Sustainability agenda will impact on the Sustainability debate at large. From a more general perspective on the latter, one could also argue that the MDGs and SDGs provide a counter-weight emphasising equity, human needs, and human well-being while environmental aspects dominate the largest area of the Sustainability debate. One important aspect behind this development certainly is that the voices of the global South are easier being heard in the context of the UN, while the broader Sustainability debate is also populated by a multiplicity of agents (including businesses) with very little interest in global redistribution.

To conclude, the current hybridity of Sustainability consists of a strong focus on the green economy, narrower environmental framings of Sustainability, and limited pockets, where equity still is the main goal. This hybrid nature is symptomatic for current understandings and practices related to Sustainability and their development since the introduction of the concept.

The three chapters that follow investigate three areas of major challenges to Sustainability in more detail. In the process, they will also take up some of the issues raised in this chapter and trace how they translate in the specific contexts of climate change, energy, and agricultural food production.

Notes

1 Genealogical work in history, philosophy, and the social sciences draws in particular on the works of Friedrich Nietzsche and, more recently, Michel Foucault.
2 See http://ec.europa.eu/europe2020/europe-2020-in-a-nutshell/priorities/index_en.htm [accessed 7 July 2015].

3

SUSTAINABILITY AND CLIMATE CHANGE

Introduction

This chapter discusses the issue of climate change as the most mediatised and heatedly debated environmental concern in the entire Sustainability debate. Whereas environmental issues like resource depletion, waste, and emissions, as well as overpopulation, were important until the mid-1980s, global warming and climate change increasingly dominated the debate afterwards. We will start off with a look at the science behind climate change and how our present knowledge about the interrelation of green house gases (GHG), industrial processes, and global warming has evolved. Special attention is paid to the hybrid nature of knowledge production within the frame set by the Intergovernmental Panel on Climate Change (IPCC). Then we focus on the unsustainability of climate change given current trajectories, not just from an environmental point of view but from its social and political consequences: climate change as an inherent socio-political phenomenon. The movement of the issue of climate change in both time and discursive space from science to politics and economics will be looked into in more detail below, with a special emphasis on the controversies around climate change that were all of a rather non-scientific nature. In the final part of this chapter, we take a look at the current and future trends in the politics of climate change and what major strategies, worldviews and ideas are struggling with each other.

Towards a science of climate change

The scientific analysis of climate change, its causes and its consequences is much older than recent debates suggest. In the early nineteenth century, French mathematician Jean Baptiste Fourier developed what he called an analytical theory of heat in which he connected intake of solar radiation and its effect on atmospheric gases to changing temperatures on the planet (Fleming, 1999). Whereas Fourier focused more on the influence of the sun and other celestial bodies, Irish physicist John Tyndall was the first to demonstrate that different atmospheric gases absorb heat to different degrees, thus discovering the chemical foundations of the greenhouse effect (Fleming, 2005). On a side note, Tyndall proclaimed that he was both a materialist as well as a pantheist, seeking to unite objective knowledge with moral and emotional nature. Being close to writers like Emerson and Fichte, he could probably be described in today's terms as a deep ecologist (Barton, 1987). In most historical accounts of the emergence of climate science, the title of *father* is usually ascribed to the Swedish scientist Svante Arrhenius, one of the founders of the field of physical chemistry. He developed a theory to explain the occurrence of the ice ages depending on the amount of carbon dioxide available in the atmosphere in 1896. He further coined the notion of a 'hot house' that successively became well-known as greenhouse effect (Rodhe, Charlson & Crawford, 1997). Arrhenius was mostly occupied with the possibility of a new cooling of the Earth's climate and concluded that if carbon dioxide levels in the atmosphere would drop to fifty per cent compared to the late nineteenth century, global surface temperatures would fall by four to five degrees Celsius. However, he also acknowledged that the flipside of his calculations implied that doubling CO_2 levels would result in a similar rise of global temperatures. Arrhenius even connected such a possibility to the increase of coal burning in industrial processes but foresaw this for a very distant future. In the following decades, Arrhenius's work was criticised with regard to the data he used and his estimates, as well as with regard to the assumptions of his theory in general.

The major scientific controversy about the global climate really took place in the first half of the twentieth century (Fleming, 2005, p. 107). From the 1950s onwards, theories on the relationships of GHG like CO_2, their emission from industrial processes and their effect on global temperatures became more sophisticated. In retrospect, Arrhenius's theories were already quite accurate as latter assessments on the basis of more comprehensive data demonstrated. Canadian physicist Gilbert Plass can be cited to having developed the best account of these inter-relations in 1956. He correctly estimated an average temperature increase of about one degree Celsius for the year 2000 compared to 1900. Although Plass himself acknowledged that this could be ascribed to mere coincidence and luck, it was later confirmed that some of his estimates had been erroneous but cancelled each other out thus producing the correct result (Schmidt, 2010). Probably just as interesting for our discussions here, Plass chose to publish his article in the rather widely read *American Scientist* and, thereby, immediately reached a broader audience.

Around the same time, Charles Keeling began to record global CO_2 concentrations at the Mauna Loa observatory in Hawaii. He had measured carbon dioxide concentrations in California and Washington earlier but it was in Hawaii where the so-called Keeling curve was established as probably the most well-known time series of CO_2 in the Earth's atmosphere (Keeling, 2008). Keeling's original insight was that concentrations changed in cyclical patterns, from day to night, and also within the change of the seasons. He realised that this was due to the respiration of plants and their changing intake of carbon dioxide. From the beginning of measurements up until today the Keeling curve shows a clear upward trend of CO_2 in the atmosphere, measured in parts per million (ppm), or how many micrograms of carbon dioxide can be found in one cubic meter of air. Starting in the late 1950s with just over 310 ppm, the peak in 2014 was well over 400 ppm.

The scientific consensus on the origins, mechanisms and effects of climate change then evolved over the following five decades thus establishing climate science as an innovative new project within science itself: an interdisciplinary research programme holistically aimed at addressing a global issue. When we talk today of inter- or multidisciplinary research, climate science was at the forefront of this approaches (Chen, Boulding & Schneider, 1983). During the 1990s and early 2000s, climate science became a central element of a newly emerging and more encompassing scientific programme termed 'Earth System Science', which also incorporated social sciences and the humanities (Houghton & Change, 1996; Mackenzie, 1998). Another interesting part of this research programme is now called Earth System Governance, which focuses on the interconnected systems of formal and informal rules, rule-making systems themselves, and networks of diverse actors from local to global that are trying to steer society towards preventing, mitigating, and adapting to environmental changes like climate change within the normative context of Sustainability (Biermann, 2014).

The pivotal organisation identified with the latest most widely accepted research into the nature and effects of climate change is the IPCC. As its name already denotes, the IPCC is not only responsible for research but a nongovernmental hybrid between climate research and global political institutions. Established in 1998 by the World Meteorological Organization and the United Nations Environment Programme (UNEP), the IPCC is equipped with a mandate by the United Nations to do in-depth assessments on climate change using the latest available data, models and scientific understanding. Its assessment reports are peer reviewed before publication in a twofold manner: first there is a scientific review from different experts on the topics reviewed; second there is a 'political' review by all IPCC member governments (with support from the initial reviewers). Thus the final reports, the so-called synthesis reports, are not a product of science but also of politics, showing how politically contested the subject of climate change has become – or as Shardul Agrawala has put it:

> [T]he IPCC plenary approval process of policymaker summaries often resembles a fox-trot performed by a drunken couple: one lurch forward, followed by a sideways stagger, then a stumble backwards … The final negotiated statements from such sessions are often based on least common denominator conclusions written in carefully hedged language.

> *(Agrawala, 1998, p. 627)*

One benefit of the IPCC process being such a scientific-political hybrid is that its end results can be seen as rather uncontested, even if they appear more conservative or less bold in the eyes of climate activists. With its synthesis reports, the IPCC has had a great influence in public and political perception and understanding of climate change and continues to do so. It created a global epistemic community for climate change and, given the diverse backgrounds of its scientific expert members as well as member governments from around the globe, this epistemic community inhabits many diverse viewpoints from both the global North as well as the South. However the IPCC has been challenged for its hegemony in the discourse and its perceived weakness due to heavy government influence. Of course, this can also be viewed as an unavoidable feature of a global multidisciplinary and heavily politicised research endeavour that is climate science. Some do in fact see the IPCC as new form of knowledge production in the twenty-first century (Hulme & Mahony, 2010).

Dimensions of unsustainability

Going back to the Keeling curve, the trajectory and scope of change within atmospheric concentration is getting clearer when matched with historical data about the Earth's climate and past CO_2 concentrations. In particular, data gathered by drilling into the Antarctic ice shield and analysis of ice-cores is opening a window into the distant past. This enables us to trace the phenomenon of climate change back to times long before homo sapiens inhabited the planet (Petit et al., 1999). Ice core drilling started in the 1950s and since the analysis of an ice core drilled at Vostok station, a Russian research station on Antarctica, the connection between rising CO_2 concentrations and global temperatures can be regarded as safely established (Barnola et al., 1987). Results from ice core data also showed that CO_2 in comparison with other GHG had had the most significant effect on climate in the past three centuries, since it became the predominant atmospheric agent emitted by human activities in the course of industrialisation (Intergovernmental Panel on Climate Change, 2007). The later Mauna Loa measurements derived by Keeling and others, combined with the historical data won from ice-cores, make it safe to argue that atmospheric CO_2 concentrations have never exceeded 300 ppm in the documented climate history of mankind. Along similar lines, William Nordhaus in 1975 stated that '[if] there were global temperatures more than 2 or 3°C above the current average temperature, this would take the climate outside of the range of observations which have been made over the last several hundred thousand years' (Nordhaus, 1975, p. 23). This temperature range is also significant when comparing it to the so-called 'Little Ice Age' from the thirteenth to the early nineteenth century, when average air temperatures were 2°C lower than today (Mann et al., 2009). During that period sea surface temperatures were colder with possibly increased sea ice extent and much stronger storms. In the Northern hemisphere this resulted in harsher winters, reduced crop yields and famines. Going back further in time, to the last greater glacial period commonly referred to as the 'Ice Age', we see a temperature difference from around -3°C (Yin & Battisti, 2001) to -4°C (Houghton et al., 2001). Global average temperature changes in these ranges cause dramatic shifts in local and global conditions with many different consequences on sea-level rise, increased damage from storms and severe weather, the food system, droughts, heat-related illnesses and diseases, as well as economic losses resulting from them. The latest synthesis report by the IPCC states:

> that limiting total human-induced warming to less than 2°C relative to the period 1861–1880 with a probability of >66% would require cumulative CO_2 emissions from all anthropogenic sources since 1870 to remain below about 2900 $GtCO_2$ … About 1900 $GtCO_2$ had already been emitted by 2011.
>
> *(Intergovernmental Panel on Climate Change, 2015, p. 10)*

Accordingly, this scenario of course requires decisive measures rather than a business-as-usual approach to GHG emissions. If the present trajectories remain in place, the report finds that a temperature increase of 4°C by the end of this century is very likely. The effects of that would be highly unsustainable. Given a temperature increase of 2°C to 4°C, it is estimated that sea-levels will rise by 0.5m and 2m by the end of the twenty-first century compared to 1990 (Nicholls et al., 2011; Rahmstorf, 2007). In Europe, this would flood significant parts of the Netherlands and the German North Sea coast, as well as the Northern Adriatic shoreline in Italy. In North America, the metropolitan areas on the Atlantic and also New Orleans would come under stress.

The estimated cost of fully protecting these areas would thereby exceed the costs of losing land – which would create a politically sensitive situation with the possibility of large scale resettling of more than 180 million people (Bosello, Roson & Tol, 2007). The impact on food was detailed as early as 1979 in the so-called Charney report (National Research Council, 1979). Its main findings were that doubling CO_2 levels from a 1970s perspective would most likely result in a temperature increase close to three degrees with severe effects on the agricultural sector by reducing amounts of arable land and crop yields, thus increasing the risks of widespread famines. These findings were validated by more recent studies stating that especially low-income producers and consumers of food will be more vulnerable to climate change (Vermeulen, Campbell & Ingram, 2012) and that climate variability and change will exacerbate food insecurity in areas currently vulnerable to hunger and under-nutrition (Wheeler & Braun, 2013). Climate change will also increase the occurrence of heat-related diseases. This will especially become a problem in densely populated urban areas both in the global North and South, driving up mortality rates for the elderly and people suffering from cardiovascular diseases, thus increasing the vulnerability of an ageing population (Luber & McGeehin, 2008). There will also be an increase in the likelihood of infectious diseases and their proliferation across the globe due to shifting climate zones (McMichael, Woodruff & Hales, 2006). German social psychologist Harald Welzer sees societies coming under pressure from climate change in danger of resorting to violence to cope with its consequences. Especially in countries with weak government institutions that are already under stress with regard to food supply, access to water, and other natural resources, the possibility for 'climate wars' increases (Welzer, 2012). There are many more consequences of climate change that add up to an increase in unsustainability across many different sectors. This section cannot spread them out in full detail but it should have become clear the disruptive potential climate change has – not in its direct ecological impact, but as social, political and economic stress. Closely connected to our perspective on the idea of Sustainability, we see climate change as a hybrid phenomenon driven by human activity and resulting in social and political struggles about the framing and understanding of its implications and meanings. This constitutes climate change as a highly contested social process.

Climate change as a discursive field

From the 1970s until today, climate change debates left the confined sphere of science and transgressed into political, economic, and popular debates resulting in an all-encompassing discursive field. In the process, the significance of climate change is not only assessed on the basis of natural-scientific data but also, for example, in political debates about global climate agreements, which are inevitably linked to geopolitical considerations. Its close alignment with the notion of Sustainability changed the way climate change is addressed today. And climate change, as part of the discourse on Sustainability, was translated into carbon-friendly or carbon-neutral ways of living, working, and doing business. In the following, we take a more detailed look how this discursive field has evolved and how developments in this field have affected notions and practices of Sustainability.

From science to global policy

The original report on the 'limits to growth' did not mention climate change. Dennis Meadows himself once stated, that climate change has to be perceived as a 'wild card' that his team was not aware of in 1972. However the discussion about climate change clearly influenced how the messages of *Limits to Growth* were perceived. Interestingly, CO_2 disposable rights (Barney, 1980), that is, emission permits as in today's European emissions trading scheme (EU ETS), were seen as a means to address climate change as far back as 1980 in the *Global 2000 Report* to US President Jimmy Carter is a very early example of how climate change made the transition from science to politics – at least in the form of policy recommendations. It must also be added that a reduction in CO_2 and other GHG emissions seemed hardly feasible back in those days. The only option considered in the *Global 2000 Report* was massive reforestation programmes and a dramatic increase of nuclear energy – a solution that Plass had already dismissed as unrealistic in the 1950s. In these early years of the Sustainability discourse, the issue of climate change was not so much formative for a new political project or climate aware lifestyles but for establishing the interdisciplinary field of climate science, an interdisciplinary scientific community concerned with climate change and global warming, and the possibilities of limiting carbon dioxide emissions in order that humanity would remain within a certain safe operating space (Weart, 2013). With the launch of *Climate Change* as an interdisciplinary journal for climate research, the field could identify itself as inter- and multidisciplinary. Researchers could acknowledge that going beyond disciplinary boundaries would be necessary for an issue that others have regarded a 'wicked problem' (Rittel & Webber, 1973) that even required a transdisciplinary approach (Brown, Harris & Russell, 2010).

The Brundtland Report and its rhetoric did not emphasise climate change as an overarching threat but rather as one environmental problem amongst many others. For example, the conservation of living natural resources, later better known as biodiversity (WCED, 1987) also received much attention. Climate change was rather perceived as a sub-theme of energy issues, especially fossil fuels – a topic that we will discuss in much greater detail in Chapter 4 of this book. Crucially, the United Nations Framework Convention on Climate Change, signed at the Earth Summit in 1992, installed a political process of climate protection and the 'Conferences of the Parties' known as COP (United Nations, 1992b).

Even though the UNFCCC did not prescribe binding goals, for example, to limit temperature increases, it paved the road to the Kyoto Protocol in 1997. The intention of the Kyoto Protocol was to formulate globally binding emission-reduction targets for GHGs for each member country to prevent dangerous consequences for the climate system. However, it was very much open how this could be achieved. The two most discussed options were market-oriented. On the one hand, specific taxation of carbon or fossil fuel; on the other hand, a cap-and-trade system with tradable emission permits (Ekins & Barker, 2001). Both instruments use the price mechanism as an incentive for a cost-efficient reduction of carbon dioxide emissions. While taxation directly puts a price on carbon, tradable emission permits put a constraint on the amount of emissions individual agents can produce, thus indirectly leading to price changes via a permit market. A carbon tax also directly creates revenues which can be used for tax reliefs in other areas, for example, for funding technological innovation or easing the tax burden for energy-intensive industries. With emission permits, only auctioneered permits create revenues; if permits are 'grandfathered', i.e. given freely to emitters based on their past emissions, no revenues are created and the market mechanisms will work slower in reducing CO_2 emissions (Singer, 2000).

Both approaches were considered as options for a global system of emissions reductions. However, it was due to political reasons that the cap-and-trade-approach was eventually chosen for the Kyoto Protocol. Especially the USA but also Russia favoured this solution, whereas European countries championed taxation. The reason behind this was that in the 1990s the US economy was still rather energy and resource inefficient compared to the EU's economy. A global taxation scheme would have resulted in competitive disadvantages for the USA, but also Japan, vis-à-vis the EU countries (Grubb, 2003). Therefore, a cap-and-trade system allowing for both compensatory measures in emerging and developing countries – the so-called clean development mechanisms (CDM) – as well as the possibility to buy emission permits from countries that do not need them was established as the political consensus. Russia showed an interest in cap-and-trade for similar reasons. After the breakdown that followed the dissolution of the Soviet Union, Russian CO_2 emissions fell significantly compared to 1990, the year used as the basis to calculate emissions reductions (Depledge, 2000).

In the process, towards the end of the 1990s climate change had become a key political project within the context of Sustainability. The COP process, the Kyoto protocol and the IPCC all stand witness to the politicisation of a once scientific question and its entanglement with the idea and discourse of Sustainability. Preventing human-made climate change became the dominant environmental concern within this broader context. Climate change at the turn of the millennium had made its transition from science to global politics.

The controversy around climate change

As already mentioned, climate change was a highly controversial scientific topic until the 1950s. A different type of controversy involving political clashes and intense mediatised debates followed when climate change became more than a purely scientific theme. Changing patterns and contents of contestation are not surprising when a discourse moves from one societal arena to another, especially if that arena is politics. The political climate change debate also encountered climate scepticism as well as outright climate change denial. Whitmarsh argues that the reasons for acceptance or refusal of climate change are intricately linked to values and worldviews (Whitmarsh, 2011). Hence, the degree of scepticism towards climate change is not a matter of knowledge about the topic itself but rather is very closely related to personal political orientation and environmental values: the further right-of-centre one's political views and the lower the value attached to the natural environment, the more sceptical of climate change people are. What is also significant is that public opinion was until recently convinced that there is a controversy amongst climate experts, despite the strong consensus reached in the field of climate science. This demonstrates that the influence of science on debates in the public-political realm is limited and that the level of contestedness might not decrease due to increasing scientific evidence. Multiple voices, conflicting values and opinions as well as a broad spectrum of media contribute to how science is received and how the issue of climate change is being dealt with (and, indeed, other issues too).

An interesting case here is the so-called 'climategate' scandal of illegally obtained and leaked e-mails from the Climatic Research Unit (CRU) at the University of East Anglia (UEA) in the UK in 2009. Within the documents made public, CRU director Phil Jones wrote to Michael Mann, the director of the Earth System Science Center at Penn State, describing a 'trick' to allegedly hide the decline in global warming over the last few decades (Leiserowitz et al., 2013; Nerlich, 2010). Michael Mann back in 1999 discovered the so-called hockey stick curve of global temperatures – the long-term decline of mean global temperatures until the twentieth century, followed by a steep increase that was supposedly following the increase in GHG emissions (Mann, Bradley & Hughes, 1999). The hockey stick curve itself rose to prominence with climate sceptics as the method and data leading to its findings were heavily criticised but ultimately supported by the National Research Council in 2006. The debate within and beyond science on the hockey stick inspired Mann to write a popular book, in the title of which he refers to 'the climate wars' (Mann, 2013). Going back to the UEA e-mails, the simple word 'trick' instantly ignited a media discussion and the notion 'climategate' emerged alluding to the

Watergate scandal in the 1970s. For both Jones and Mann this resulted in critical reviews by their respective institutions, but it was by far not restricted to them. Also the IPCC was alleged to have ignored dissenting articles while composing its fourth *Assessment Report*. As several errors were found within this report, the United Nations ordered an independent review of the internal review processes of the IPCC. However, the heart of the matter was much more trivial from a scientific perspective as the word 'trick' referred to a mathematical procedure of how to make a moving average of a time series of data. The inquiries set up to investigate climategate then found that there actually was no scientific climategate: the data and the findings were by and large correct (Biello, 2010).

In recent years another assumed controversy has reigned in the media regarding climate change. This time it was about the 'global warming hiatus'; that from 1998 onwards average global temperatures did not increase as expected but in fact stalled. As climate scientists are always pointing out, climate is unlike weather, it is a long-term trend phenomenon but after several years of this hiatus, climate sceptics like the UK-based non-peer reviewed journal *New Scientist* – who famously ran a cover in 2009 headlining 'Darwin was wrong' – picked up on it and once again declared climate change as a 'myth' (*New Scientist*, 2007). However, as recent studies show the hiatus measured was in fact more of a measuring phenomenon than a climate phenomenon, with time series showing clearly a non-disturbed increase of global temperatures during the period of 2000 to 2014 (Tollefson, 2015). However, as has been argued above, the contestedness of climate change is not about its scientific validity but about its political relevance and how it connects to existing worldviews and value systems. As the core element of the environmental perspective of Sustainability, climate change can only be understood in its connection to the wider socio-political normative frame of sustainable development that encompasses and pays attention to all aspects of society. In conclusion, the science behind climate change is rather clear about human-made carbon dioxide emissions being the predominant reason behind rising global temperatures and possible catastrophic socio-economic consequences (Cook et al., 2013) – but as part of Sustainability, climate change belongs to a 'post-normal' situation where uncertainties in understanding and outcomes as well as political stakes are high and new knowledge production will always remain contested (Ravetz, 2011).

Towards the dominance of climate change

Looking at the broader Sustainability debate from a perspective on climate change, the controversies about the empirical evidence for climate change, the role of CO_2, or human activities were not the most significant developments. The controversies were most intense in a phase, which we described as a phase of transition with regard to the dominant meanings and activities related to Sustainability. The strong focus on global equity the needs of the world's poorest people as advocated in the report *Our Common Future* did not prevail for long after its publication. Instead, global debates about Sustainability shifted towards a much stronger economic framing.

The British government commissioned a review on the *Economics of Climate Change* in 2006. Its author, Nicholas Stern from the London School of Economics and Political Science, argued from an economic perspective that preventing climate change as early as possible would not only pay off ecologically but also economically (Stern, 2007). Prevention measures, so the Stern Review reasoned, would act as a giant green investment project boosting economic growth and creating new jobs in new industries. We have already seen in Chapter 2 that Thomas L. Friedman picked up this line of thinking and called for a Green New Deal, understood as a great industrial project. In 2008, the UN formulated their Green Economy Initiative in a very similar key (Borel-Saladin & Turok, 2013; Friedman, 2007). These economic assessments and calls to action contributed to making climate change the central environmental project with regard to Sustainability. Moreover, they introduced economic renewal in the global North and ecologically sound economic development in the global South as clear reference frames. This does not imply that climate change had ceased to be politically contested. The movement of the issue of climate change from science to politics to economics shows, however, how a project with relevance for society as a whole is shaped. In this process the reference frame of climate change broadened, came into conflict with worldviews and value systems, and established connections between other sustainability-related fields, namely energy (Chapter 4) and food (Chapter 5). All fields of Sustainability now comprise of a more and more solid nexus of cross-references and stabilise each other – and therefore ensure the coherence of the Sustainability discourse as a whole.

Climate change between mitigation and adaptation

At the time of writing this book in 2015, it seems safe to argue that the scientific facts gathered on anthropogenic climate change are regarded across most of the political spectrum as a form of 'truth' (Thee-Brenan, 2014), though ensuing debates about the political consequences show that this issue is still far from being solved. In other words, the proper way of dealing with climate change is still in doubt and contested – at times quite heavily. This holds for politics as well as for individuals. Possible reactions can be broadly classified as mitigation or adaptation. Mitigation aims to reduce GHG emissions in order to prevent climate change from happening, at least beyond the two degrees' guardrail. Adaptation is based on the assumption that climate change is happening and that human communities need to adjust to these changing conditions. It refers to reworking humanity's infrastructures and political systems to deal with the consequences of climate change. While the main focus of previous global climate policy, in particular under the UNFCCC, was on globally coordinated efforts to mitigate climate change, adaptation recently won more credibility as a viable political alternative (King, 2004; Pielke et al., 2007). Adaptation allows for local, individual responses in the place of general global approaches to climate change; this has become ever more important since negotiations to find another agreement to succeed the Kyoto Protocol have repeatedly stalled.

Within the continuum of mitigation and adaptation we can detect two further ideal-typical patterns of understandings and worldviews, of legitimation and value, against which contested projects and concepts are checked and evaluated. The first one can be described as *technological voluntarism*. This means that technological solutions are at the core of most attempts to address climate change. This is combined (or motivated by) the vision of minimal intervention in markets, orders, and individual lifestyles. In the context of this first pattern, mitigation translates into ecological modernisation and the idea of green growth (Buttel, 2000;

Hallegatte et al., 2012). Free markets and setting the right incentives for environmental and climate-friendly technologies are seen as enough and it is believed that climate change can be (or even should be) tackled on the basis of free-market liberalism, capitalism and some mild government interventions. Such a position assumes that absolute decoupling of economic growth from increased ecological impact is possible.[1] With regard to adaptation to climate change this voluntarist position can take various forms; the most prominent one is that of geo engineering or climate engineering. Climate engineering embraces the active human influence on natural systems and advocates not less human footprint but more. Its supporters promote technologies like removing carbon dioxide from the air via sequestering technologies to store it (carbon capture and storage), for example, underground or on the sea floor. Another, still hypothetical example is the plan to undertake solar radiation management and to curb global warming by deflecting sunlight from the Earth's atmosphere (Bengtsson, 2006; Blackstock et al., 2009). Proposals of this kind range from utilising space based mirrors to spraying aerosols into the upper atmosphere or seawater into the air to create more clouds, which should reflect sunlight away from the earth.

The second pattern can be described as *sufficiency*. Sufficiency means self-restrain as regards production and consumption in climate-intensive supply chain logistics, including self-sufficient production of goods and services (Princen, 2003, 2005). For mitigation of climate change, sufficiency translates into transforming lifestyles to become less carbon intensive and production to move closer to consumers and their life worlds. In practice, the patterns can overlap and the ideal typical distinction can be blurred. For example, the idea of sustainable consumption promoting eco-friendly products and greening of the supply chain (Humphery, 2015) appeals in both contexts. However, sufficiency opposes the growth-oriented expansion of free-market capitalism, instead promoting a view on mitigation that is based on reduction and contraction. This view assumes that absolute decoupling of economic growth and ecological impact is impossible. Importantly, sufficiency does not equate with individual austerity but instead bears a more political message of specifying the boundaries of production and consumption within hard ecological limits (Alcott, 2008). In other words, it draws on a notion of Sustainability that is influenced by that of an equilibrium put forward by the authors of *Limits to Growth* and translates this idea into a critique of entirely unregulated capitalist consumer societies – similar to the Brundtland Report, which framed growth as a means to provide well-being to those in need.

With regard to adaptation to climate change, sufficiency leads to the idea of resilience. Resilience is understood as the capacity of any kind of system to uphold and regain its functioning after severe shocks (Holling, 1973; Walker & Salt, 2012). Comparing the consequences of hurricane Katrina in New Orleans in 2005 and that of hurricane Sandy in New York City in 2012 provides a good illustration of this issue (Nemeth, 2013). Following Katrina, neither the City of New Orleans, the state of Louisiana, nor federal assistance could undo the damage and recreate stable and viable living conditions, which amounted to systemic failure. As a system, these actors failed in providing for resilience. The situation after Sandy was totally different. The subway system went out of operation only for a few days with most of New York's public infrastructure services functioning again shortly afterwards. Obviously, some lessons were learned from Katrina playing into the hands of this particular system and overall the reaction to Sandy showed much more resilience. Risk management, flood management, food management, and so on, are part of resilience as a project of sufficient adaptation to climate change. Recently, the concept of sufficiency and its inherent messages of reduced lifestyles and resilience has gained more prominence.

A good example for a sufficiency oriented response to climate change is the rise of the so-called 'degrowth' movement (Demaria et al., 2013; Latouche, 2010). Degrowth implies a voluntary and planned transformation of political and economic systems to a non-expansionist society that is not depending on growth. In turn, it aims for a cultural change away from consumer capitalism, for refocusing politics not on economics but people's lives and well-being, as well as on caring for planetary ecosystems within hard ecological boundaries. The Simplicity Institute in Australia has taken the degrowth idea to the question of preventing climate change. It calculated that technological voluntarist approaches would not be enough but needed to be complemented by some form of controlled economic contraction (Alexander, 2014).

To conclude, we can see the emergence of two ideal typical patterns within the political-economical reference frame of climate change – technological voluntarism and sufficiency. Moreover, we detected three larger responses to climate change, in this context: one majority project of eco-modernism, green growth and a decarbonized world economy, which is based on the hope that economic growth can be decoupled from material growth in terms of resource consumption, waste, and emissions. This framework has become the preferred route to sustainable development in general. In addition, we described two minority projects. On the one hand, the technological project of climate engineering; on the other hand, the sufficiency project of resilience and degrowth. Whether one of the minority contenders can break into the mainstream rests also on their ability to enter new societal fields and discursive spaces. For example, degrowth would have to become more than just a research agenda or an activist movement; it would have to become a political issue and a business matter, for example. Similarly, geoengineering would have to leave its hypothetical scientific niche, and be translated into political-economic projects within the wider frame of Sustainability. Until now it seems more likely that geoengineering will become a part of the mainstream within the green economy narrative, rather than overthrowing it or replacing it (OECD, 2011).

Note

1 While relative decoupling means reducing additional ecological impact per additional unit of economic growth, for example making a combustion engine run on less fuel, absolute decoupling demands a decrease of total ecological impact per additional unit of economic growth, implying to leave the natural environment more sound after consuming it than before.

4

SUSTAINABILITY AND ENERGY SYSTEMS

Introduction

This chapter discusses Sustainability in the context of energy systems. The permanent availability of relatively cheap and abundant energy is a key characteristic of modern industrialised societies. At the same time, this backbone of modern societies has massive social and environmental consequences not only because incumbent energy systems are to a significant extent based on fossil and nuclear energy sources. The problem of climate change is only the most prominent Sustainability aspect in this context. Therefore, many see the transformation of energy systems as key for greater Sustainability.

The chapter starts by outlining the formation and key characteristics of contemporary energy systems. Then, it points to several problems of unsustainability based on incumbent practices of energy production and consumption. Due to the complexity of energy systems and due to their ubiquity in modern life, a comprehensive discussion of all related environmental and social problems would be beyond the scope of this book, if it were possible at all. Therefore, we outline only a selection of significant issues to provide an impression of the larger picture. Moreover, particular unsustainability problems can be related to various specific responses, which can, for example, emphasise different aspects and analyses of the problem, prioritise different strategies and activities, and draw on different understandings of Sustainability in the field of energy. Therefore, it is equally impossible to present the multiple understandings and practices of Sustainability in their entirety. Instead, the following discussion focuses on selected projects of Sustainability with regard to energy consumption and production in order to explore how this idea operates in this field. Particular attention is paid to the larger patterns outlined in Chapter 2, which will be traced throughout the chapter and contrasted with the exemplary projects of renewable energies, the quest for equal access to energy for all and finally the business case of sustainable energies. The discussion elucidates that Sustainability, understood as a precept to respect ecological limits, can at least implicitly be observed in projects that emerged around the utilisation of renewable energies. While equity has been emphasised in the understanding of Sustainability formulated in the Brundtland Report and is still an important value in various projects, recent approaches to sustainable energy are set on the understanding of Sustainability as a business case, seeking to balance environmental and economic benefits.

The formation of modern energy systems

In order to better understand the Sustainability challenges arising in contemporary energy systems, we begin this section by briefly reconstructing their historical formation. Energy was always essential for the constitution of human societies. For a long time, the utilisation of energy was limited to relatively simple practices of production and consumption (i.e. local, decentralised, and not requiring the collaboration of specifically skilled people or extensive technological infrastructures) and renewable energy sources. For instance, firewood was burned for heating and cooking, mills and wheels were driven by the power of wind or water, and transportation often rested upon the power of horses or men (see, Sørensen, 1991).[1]

A new chapter was opened in the eighteenth century when coal was first used in steam engines. (This brief historic review draws on Union of Concerned Scientists, 2015.) Subsequently coal became an important fossil energy source fuelling steam engines to run railways and steam ships or to power factories and smelting furnaces for steel production. From now on, the energy mix was gradually shifting towards fossil energy sources. This development gained pace when crude oil as another fossil energy source came up in the late nineteenth century. Initially oil was merely used for minor purposes, but soon considerable amounts were processed and used in combustion engines. Their evolution set the scene for new developments, such as the mass production of cars and the extension of car-centred mobility infrastructures. In addition to transportation, crude oil became the foundation of a variety of new practices and infrastructures, including heating and warm-water systems or various industrial production processes. Also during the late 1800s, electricity systems including power plants and distribution networks emerged and spread quickly. Finally, in the 1950s nuclear technology was applied to develop a new opportunity to produce electric energy. (For a detailed historic account also see Smil, 2004b.)

The rapidly growing production and distribution infrastructures for fossil energies (coal, oil and later natural gas) and electricity (partially based on fossil fuels and nuclear power) provided more and more people in industrialised countries with relatively cheap energy at any time and in most places. Fossil energy could be provided at costs much lower than renewable sources, for instance in '1900 solar power was estimated to cost about 10 times that of the competing fossil power' (Sørensen, 1991, p. 10). In the process, widely and abundantly available energy became a key characteristic of modern industrialised high-energy societies. In today's energy systems fossil fuels account for more than 80% of the world's total primary energy supply including coal (29%), oil (31.4%) and natural gas (21.3%) – nuclear power plays an important role in the provision of electricity with a share of almost 11% of total electricity generation (data for the year 2012, International Energy Agency, 2014).

Alongside the advancements in energy provision, an increasing number of energy consuming devices, from cars to refrigerators, quickly spread throughout everyday life at work and in private homes. In the process, industrialisation and new energy consuming lifestyles led to an explosion of energy use in absolute terms as well as per capita (Beretta, 2007). Accordingly, the production, distribution, and consumption of energy has become deeply embedded in most modern practices of everyday life. Today's societies would look indeed very different if the world were to run out of power. In fact, large power failures resulting from bad weather or grid problems demonstrate with some regularity how deeply contemporary urbanised societies are embedded in energy. One example is the so-called Northeast blackout of 2003, which seemed to be caused by a software bug and affected 50 million people for days in the USA and Canada. In the USA the costs caused by the outage were estimated at between four and ten billion US dollars, while in Ontario the monthly GDP dropped by 0.7% and a loss of almost 19 million work hours was estimated (US–Canada Power System Outage Task Force, 2004). Another example is the European blackout of November 2006, which once more illustrated the vulnerability of highly interconnected energy infrastructures. The blackout that finally affected 15 million households in 20 European countries started by a poorly planned shutdown of a transmission line in Germany, after which 'in an astounding 14 seconds a cascade of power line trippings spread through Germany. In the next 5 seconds the failure cascaded as far as Romania to the East, Croatia to the South-East, and Portugal to the South-West' (van der Vleuten & Lagendijk, 2010, p. 2043). Such outages do not only leave people behind in dark homes or stuck in elevators but may cause issues in critical infrastructures such as transportation (especially railways), telecommunication, water supply or hospitals.[2] This is also the reason why blackout scenarios work as immensely powerful threats in debates about energy transformations in general, and renewable energy based on wind and solar in particular. Because energy production from wind turbines or photovoltaic panels is not steady but volatile they may challenge system stability. Therefore the risk of power outages due to network instabilities is often raised as an argument, for instance by industry representatives, to limit or slow down the integration of renewable energy production capacities into existing systems.[3]

Dimensions of unsustainability

While modern energy systems have obvious benefits, they also produce considerable environmental and social effects. Therefore, modern energy systems are extremely ambivalent from a Sustainability perspective, mainly but not only due to their reliance on fossil-nuclear fuels and technologies. Resulting GHG emissions contributing to climate change (see Chapter 3) are the most prominent concerns in recent debates and calls for more sustainable energy systems. For example, the Intergovernmental Panel on Climate Change (IPCC) estimated that the 'consumption of fossil fuels accounts for the majority of global anthropogenic GHG emissions' (IPCC, 2011, p. 7). In addition, other pollutants such as sulphur or particulates are emitted when fossil fuels are burned, posing threats to human health or causing environmental degradation, for example, through acid rain. Furthermore, mining and extraction can devastate considerable areas, put pressure on local ecosystems, and endanger inhabitants' well-being not only by consuming land that could be used otherwise (for example, for farming or housing) but also by releasing problematic substances and polluting soil, air, and water.[4]

This brief sketch shall indicate that incumbent energy systems continuously overuse ecological capacities (for instance GHG sinks) and are thus unsustainable in multiple environmental dimensions. Given the primary concern of climate change, nuclear power is sometimes presented as a carbon neutral alternative to fossil fuels. For example, the UK Government has declared nuclear power a core element of both its industrial policy as well as its attempts to curb GHG emissions (see HM Government, 2013a, 2013b). However, nuclear power bears the additional and immense risk of disastrous accidents as incidents in Three Mile Island, Chernobyl, and Fukushima demonstrated dramatically. Moreover, there still is no solution to the problem of nuclear waste, nor to the problem that nuclear material can also be used for weapons.

In addition to the limitations imposed by the environment both fossil and nuclear energy sources are non-renewable and thus finite. Hence, at increasing rates of consumption the depletion of primary sources is just a matter of time (see the discussion on peak oil below). The finiteness of non-renewable energy sources in addition to the political instability of major extraction regions may not only cause energy prices to increase but also raise serious concerns of energy security, as it might not be possible to maintain the current supply in the long run.

Moreover, fossil-nuclear energy systems are mainly based on centralised, large-scale power plants as well as on extensive distribution infrastructures. At the same time, energy markets are often characterised by a concentration of power. Only a small number of multi-national corporations control the majority of global energy supply from fossil and nuclear sources. In addition, due to high investment costs and technological complexity, access and scope for participation for a diversity of actors is limited.

In recent years, increased costs of exploitation and extraction as well as the necessary high investments from fossil fuel companies were criticised as a 'carbon bubble' (Carbon Tracker Initiative, 2012). Borrowing the bubble terminology of unfounded stock market or real estate market exuberance, the carbon bubble is described as the difference between the potential release of CO_2 presently stored in listed fossil fuel reserves and the remaining global CO_2 budget left to prevent overshooting the 2°C guardrail for climate change. The scientific foundation of the carbon bubble was formulated in 2009 by German climate researcher Malte Meinshausen and his colleagues. Their findings show that more than half of the listed fossil fuel reserves would need to remain in the ground in order to achieve current climate change goals – hence the term 'unburnable carbon' (Meinshausen et al., 2009). If these climate change goals remain on the political agenda without major revisions, all fossil fuel companies would be overvalued and in danger of facing a dramatic re-evaluation on the stock markets.

Finally, access to energy is distributed highly unequally between the global North and South. For example, in a special early excerpt of the World Energy Outlook 2010 for the UN General Assembly on the Millennium Development Goals the International Energy Agency (IEA) draws attention to the problem of energy poverty as an often neglected issue in the quest for sustainable energy solutions:

> It is the alarming fact that today billions of people lack access to the most basic energy services, electricity and clean cooking facilities, and, worse, this situation is set to change very little over the next 20 years, actually deteriorating in some respects. This is shameful and unacceptable.
>
> *(International Energy Agency, 2010a, p. 8)*

Taken together, the mentioned concentration of market power, economic bubbles, and especially the global inequalities regarding the access to energy additionally undermine the Sustainability of incumbent energy systems. These issues require similar attention in the search for alternatives, just as GHG emissions, consumption levels or supply security.

In conclusion, contemporary fossil-nuclear energy systems are unsustainable with regard to several problems: lacking ecological viability, limited availability of primary resources, unreliable supply security, high economic vulnerability, unequal access to energy, and social costs imposed on some parts of the population, for example, on those affected by mining. Several more unsustainability challenges could be added to this list. However, the ambivalence of contemporary energy systems, which provide modern industrialised societies with cheap and reliable energy while causing entirely new ecological and social problems should be clear by now. Over time, various responses to these particular unsustainability problems of current energy systems have emerged. The remainder of the chapter discusses a selection of such responses from a Sustainability perspective.

Renewable energies as a response to finite resources

Concerns, particularly about the natural limitations of fossil energy systems, can be traced back long before the key idea of Sustainability was formulated. For example, the discussion about peak oil (see below) is a good illustration of how concerns about the overconsumption of resources triggered the search for alternative pathways towards less unsustainable energy systems.

Limits to energy resources

Non-renewable energy sources are by definition finite. By principle, they will be exhausted at a certain point in time depending on the total availability and the rate of exploitation. William Stanley Jevons (1865) was among the first voices warning that fossil energy sources were limited. In his book *The Coal Question* he was concerned with the exhaustion of coalmines as he claimed that a more economical use of coal would not reduce absolute consumption but instead increase demand. This phenomenon is known as the Jevons Paradox (see Alcott, 2005). With regard to crude oil the issue of physical limits was raised for the first time in the 1950s by geologist Marion King Hubbert, lending his name to the bell-shaped production curve of crude oil (Deffeyes, 2008). Hubbert developed a model that assumed an increasing rate of oil production, which should finally reach a peak (i.e. peak oil) and irreversibly decline afterwards. The analysis of peak oil does not imply that production suddenly comes to an end but rather means a continuous decrease in production rates. In this context, the model predicted that US oil production would reach its peak in the early 1970s. 'Almost everyone, inside and outside the oil industry, rejected Hubbert's analysis. The controversy raged until 1970, when the U.S. production of crude oil started to fall. Hubbert was right' (Deffeyes, 2008, p. 1). Later the Hubbert curve was applied to the world oil production but predictions of global peak oil remain highly uncertain and contested. For instance, the IEA stated in its 2010 World Energy Outlook that the output of crude oil would reach 'an undulating plateau of around 68–69 mb/d by 2020, but never regain its all-time peak of 70 mb/d reached in 2006' (International Energy Agency, 2010b, p. 48). Moreover, the production of unconventional oil, including oil sands and oil shale, was estimated to increase significantly. In fact, more recently, new technologies allow extracting oil and gas from shale formations (mainly by hydraulic fracturing) on a large scale. On the one hand these new extraction technologies also bear the risk of considerable environmental side effects, mainly due to the eventual release of toxic chemicals that may for instance cause contamination of drinking water or negatively affect local air quality (for example, see R. B. Jackson et al., 2014). On the other hand, however, fracking allowed the oil production in the USA to grow substantially in recent years. The contribution of more than four million barrels to the daily production from fracking almost doubled the total US oil production.[5] As a consequence, assessments of peak oil, including those of the IEA, were revised and accordingly shifted into the future. In addition to future production scenarios, the expected development of demand for fossil fuels is crucial for supply security. Will demand for oil continue to increase or decline at any time soon due to increased energy efficiency and growing shares of renewable energy sources? Some commentators, including the IEA, already suppose that peak demand will eventually take place even before peak oil.[6] Independent of the precise calculations and scenarios however, the physical limitations in the availability of fossil fuels remain a major concern from a Sustainability perspective. The oil crisis in the years of 1973 and 1974 illustrated the immense dependency on imports of fossil fuels and left a vivid impression on how seriously fossil based energy systems may be disrupted as soon as supply of cheap fuel gets restricted. These considerations about the finite nature of fossil resources, but also the uncertainties involved in assessing the same limitations, are of course not limited to crude oil but similarly apply to coal, natural gas or oil sands.

In addition to the availability of and the demand for resources, another measure is crucial for assessing the sustainability of energy systems: the amount of net energy available or energy return on energy invested (EROI). EROI implies that energy production requires energy input, for example, for extracting, transporting, and processing raw materials, setting up power plants and distribution networks, their maintenance, and eventual dismantling (the latter is particularly important with regard to nuclear plants) as well as line and conversion losses. The resulting return is the net energy available for use. Calculating the EROI values is also affected by uncertainties since, for example, it is often difficult to establish what is counted as investment required to get energy (Hall, Lambert & Balogh, 2014; Murphy & Hall, 2011). Nevertheless, there is a robust trend in EROI values that shows a constant decline concerning fossil energy sources, especially when looking at crude oil (presented figures draw on Hall & Day Jr, 2009). In the 1930s when large-scale oil extraction began in Texas, the EROI of oil was around 100:1 – for one unit of energy put into extraction there were 100 units of energy as a gain, leaving the available net energy at 99. Over the years, as exploration got technologically more difficult – despite the progress in extraction technologies – EROI values went down to a ratio of roughly 14:1 or below. Net energy of oil-based energy systems shrunk by about 85 per cent in the last 80 years.

The trend also holds for gas including unconventional gas from hydraulic fracturing. Especially fracking has proven to be energetically not very efficient with an EROI of between 12:1 and 8:1 (Yaritani & Matsushima, 2014). Nuclear is also regarded as an energy source with a rather low EROI value. Some even argue that the values are negative, i.e. that more energy is used in the nuclear energy system to generate useable energy (Hall & Day Jr, 2009). Switching to a more renewables oriented energy systems might not at first change the problem of declining EROI, as many renewable energy sources have low EROI values as well. However, this is partly a technological problem with increasing the lifetime of, for example, wind farms or geothermal plants. Having a wind farm with an operational lifetime of 10 years delivers a net energy value of around 12 to 13 while the same wind farm with twice the lifetime would increase it to more than 21 (Atlason & Unnthorsson, 2014).

The idea of the limits imposed on energy use depending on the energy source was also discussed in the original *Limits to Growth* report from 1972 (see also Chapter 2). For example, the authors cautiously argue that most energy sources will have the side effect of emitting excess heat that will warm the Earth's atmosphere. This thermal pollution could create so-called heat islands, for example, in densely populated urban areas, or rivers through 'thermal pollution', altering the living conditions and viability of aquatic life forms. The only thermal non-polluting energy sources are solar, wind, and water – the latter two ultimately also driven by solar energy (Meadows et al., 1972, p. 73).

Nuclear energy plays a special role in the report which acknowledges that nuclear energy might provide responses to the problem of GHG emissions. However, it also mentions the complex risks and problems, most of which were already mentioned above in this chapter: the radioactive waste problem, the issue of thermal pollution, and the energy and resource intensive supply chains needed for creating the nuclear fuel rods – not to mention the global security issues involved in a large scale application of the nuclear option (Meadows et al., 1972, p. 133). The conclusion drawn in 1972 for a sustainable energy source for the entire planet was an increased investment in and harnessing of incident solar energy as the most pollutant-free energy source available to humankind (Meadows et al., 1972, p. 177).

Renewable energy without limits

As a response to the obvious impediments in sustaining the resource base of fossil energies and to the risks of nuclear power, renewable energy sources have gained increasing attention. Renewables are diverse and utilise different energy sources such as water (hydropower), wind, various types of biomass and finally solar radiation itself (photovoltaic or solar heating). Renewable energy sources are replenished by natural processes and most of them rely directly or indirectly on the power of the sun. In addition, geothermal energy is considered to be renewable, utilising the thermal energy produced within the Earth.

While we have to omit many aspects and complexities of renewable energy, (see Groß & Mautz, 2015; Nelson, 2011) it is important to note that all of them are inexhaustible as long as they are used at a rate lower than the rate of replenishment. For this reason renewable energy sources are often framed as sustainable energies, in the sense that supply can be sustained in the long run. At the same time, the IPCC has described shifting from fossil to renewable energy sources as an important strategy to reduce GHG emissions and as 'having a large potential to mitigate climate change' (IPCC, 2011, p. 7).

Renewable energy sources are not only framed as inexhaustible and almost carbon neutral, responding to the limitations of natural reserves and sinks. They are also often seen as generally clean, locally available, and less prone to instigate conflicts and power asymmetries. For instance when the European Association for Renewable Energy (EUROSOLAR) was founded in 1988, the vision was to substitute 'fossil fuels and nuclear energy systems with a sustainable, peaceful and renewable energy-based supply, founded on decentralized local resources close to the citizens' (Samuel, 2013, p. 5).

While renewable energies are closely linked to an understanding of Sustainability that seeks to overcome natural limits, additional perspectives have appeared to be relevant with regard to the unsustainability of incumbent energy systems. A key issue is the decentralised production of energy, based on local resources and controlled by local communities. Decentralised, grassroots and small-scale energy production has become a matter of Sustainability in its own right, although these aspects are often closely tied to renewables: 'Public utility companies, SMEs, municipalities and community groups are the key players in the expansion of renewable energy in the regions. Such a decentralised and seminal energy supply increases local added value, creates jobs and contributes to a sustainable energy supply' (Samuel, 2013, p. 93). As the quote illustrates, renewable energy technologies are often associated with additional positive side effects such as the creation of new jobs and economic growth.

Despite all the benefits in terms of Sustainability, renewable energy sources of course have their own challenges, for instance related to their relative cost effectiveness or their dependence on local conditions, which does not only result in volatile supply (see above) but can imply that simply not enough energy is available at a certain time pointing directly to questions of storage and transmission. Moreover, renewable energy also requires new infrastructures and interferences in the natural environment, for example, when wind farms or new transmission lines need to be built. Therefore, they are also criticised by citizens' initiatives and NIMBY (Not In My Backyard) protests. A further example, which we only mention in passing, concerns the production of biomass for energetic use, so-called bio fuels, which can have negative effects on food production and small scale farming. To conclude, renewable energy potentially constitutes a response to multiple challenges of unsustainability and 'may, if implemented properly, contribute to social and economic development, energy access, a secure energy supply, and reducing negative impacts on the environment and health' (IPCC, 2011, p. 7). As a practical response renewable energies often reflect an understanding of Sustainability that relates to dealings with natural limits and care for the earth. However, many renewable energy projects are still contested and many questions remain open on how to put renewable energies into practice on a large scale.

Safe energy for development

When the Brundtland Report brought forward its explicit and influential definition of Sustainability, it was also applied to the energy systems. The understanding of limited natural resources, present in renewable energy, was thereby broadened and now includes aspects of safety and equal access to allow development and meeting human needs. To be truly sustainable, the report states, energy sources need to be dependable, safe and environmentally sound (WCED, 1987). It continued by outlining four key criteria of a sustainable energy system: first, sufficient growth potential of energy supply to meet the needs of those whose well-being needs to be improved and of a growing future population; second, efficient use of energy with a special emphasis on minimising waste of primary resources in extraction and supply chains; third, ensuring public health by minimising or eliminating the risks to safety inherent in different sources of energy; finally, protection of the biosphere and pollution prevention (WCED, 1987). More generally, the report assessed contemporary patterns of energy use as unsustainable and the transition to a more sustainable global energy system remained unclear. This position is more defensive than in *Limits to Growth* with its advocacy of solar power. The connection of energy systems to climate change (see Chapter 3) was clear from the outset. The perspective on nuclear was a very cautious one with an interesting emphasis not on the health risks but on the cost risks. Rising costs in the nuclear industry, the report goes, have significantly reduced the earlier cost advantage of nuclear power and perhaps lost it altogether (WCED, 1987 part II, chapter 7, no. 46).

The Brundtland Report clearly called for a low-energy future, an energy development path that would focus on decoupling energy consumption from GDP growth. Energy consumption and thus potentials for savings are distributed highly unequally on a global scale. The authors highlighted that an 'average person in an industrial market economy uses more than 80 times as much energy as someone in sub-Saharan Africa' (WCED, 1987, no. 58). Therefore, they also argued that energy use in industrial societies needed to be significantly reduced to give the world's poorest the chance to develop.

The modernisation of energy systems was considered a necessary condition in order to meet this challenge. In fact, since the 1970s, significant improvements have been made, in particular with regard to energy use in buildings, transportation, households, and especially in industrial processes.[7] Today, enhancing energy efficiency remains a high priority, especially in industrial nations, in order to reduce growth of energy demand, the depletion of primary resources, and to minimise waste. While in the vision of a low-energy future energy efficiency was not limited to economic and ecological rationales but also presented as a precondition for global equity, aspects of energy costs and supply security have nowadays gained importance. This position is reflected in various policy strategies, acknowledging that energy efficiency 'has a fundamental role to play in the transition towards a more competitive, secure and sustainable energy system (...). While energy powers our societies and economies, future growth must be driven with less energy and lower costs' (European Commission, 2014b, p. 2). While energy efficiency appears as a key to multiple sustainability challenges, doubts have been formulated whether decoupling energy use from growth of GDP by technological means will be possible with regard to the necessary magnitude and time frame. Substantially reducing absolute energy consumption would require continuous improvements of energy efficiency at a rate well above the growth rate of GDP. A challenge that is further complicated by so-called rebound effects, the principle that underlies the Jevons Paradox presented above. Rebound effects refer to situations in which a reduction of energy use (as a result of increased energy efficiency) induces an impetus of new growth, in turn triggering additional energy consumption. For instance: reducing the energy consumption through better building insulation lowers energy bills. In turn the monetary savings might be spent on additional consumption, such as a holiday trip, that might even be more energy intensive.

A more critical position argues that decoupling energy consumption from economic growth might be an almost impossible task. For instance, a report of the UK Sustainable Development Commission (SDC) written by Tim Jackson (2009) stated that absolute decoupling might be a myth.[8] The report argued that, while it was easy to find historic evidence for relative decoupling, evidence for 'overall reductions in resource throughput (absolute decoupling) is much harder to find. The improvements in energy (and carbon) intensity (…) were offset by increases in the scale of economic activity over the same period' (T. Jackson, 2009, p. 8). As a consequence it is argued that strategies aiming at absolute reductions of energy consumption need to consider the limitation of economic growth itself.

Jackson's ideas as well as the more general vision of a low-energy future outlined in the Brundtland Report mirror the ideas of a whole social movement, which evolved around the idea that limiting growth – or degrowth – would be a necessary condition for Sustainability. In contrast, the degrowth movement challenges the idea of continuous economic growth for Western, industrialised nations as a necessity for development and ecological viability (Demaria et al., 2013). Yet, the objectives and strategies of degrowth are far from prominent as inherent elements of Sustainability in Western societies. Although many citizens experience being decoupled from growth, especially in the recent chain of financial crises.

However, to achieve a low-energy future with reduced absolute energy consumption in industrialised countries remains a crucial aspect of Sustainability. If energy consumption were to remain generally lower, equal access to safe energy for all might be much easier to achieve. This understanding is manifested again in more recent activities of the United Nations (UN) which reinforced the strong element of equity within Sustainability. For example, the UN initiative Sustainable Energy for All aims to address two major energy challenges alongside each other: 'One is related to energy access. Nearly one person in five on the planet still lacks access to electricity (...). Where modern energy services are plentiful, the problem is different – waste and pollution'.[9] Again aspects of equity and environmental protection are tied together under the label of Sustainability. In another statement, UN Secretary General Ban Ki-moon then linked the social and ecological dimensions back to the idea of economic prosperity when stating that energy 'is the golden thread that connects economic growth, increased social equity, and an environment that allows the world to thrive'.[10]

The business case for sustainable energy

The report *Our Common Future* also suggested a tight link between social and environmental dimensions of sustainable energy systems and economic growth. Thereby, the latter became a precondition for the former two rather than a self-contained objective (see also Chapter 2). More recently however, policy debates increasingly prioritise the economic potentials of sustainable energy within and beyond the context of development. At the same time, the discussion about energy and Sustainability understood as green growth also seems to come back to its very beginnings and the original *Limits to Growth* report with its clear advocacy of solar power. One important difference remains: *Limits to Growth* focused on scientific, in particular, physical analysis; in contrast, the green growth strategy is a cost-oriented one based on an economic analysis. In a way this mirrors the development of climate change as shown in Chapter 3, which moved across scientific, political, and economic framings of Sustainability.

As the key idea of Sustainability is increasingly narrowed to its economic dimension, sustainable energy is commonly understood as an opportunity or even a necessary condition for economic growth. Especially renewable energy technologies are often linked to the potential growth of added value, economic progress and the creation of new jobs. In the policy arena these arguments are referred to in order to legitimise public funding of research and development, as well as subsidy schemes for market implementation.

Attempts to master a notable energy transformation in Germany provide good examples.[11] The German Renewable Energies Act from the year 2000 introduced a feed in tariff for smaller producers (a large number using solar panels) to foster the extension of renewable energy. This feed in tariff and related energy policies were not only framed as a way to increase the share of renewable energy and reduce GHG emissions but were at the same time understood as economic policies (as the UK nuclear strategy mentioned above is at the same time an industrial and a climate change policy). Based on the understanding that 'renewable energy and energy efficiency are key markets to watch[, and that] German companies already occupy a leading position in these fields worldwide' (Federal Ministry of Economics and Technology, 2012, p. 2) the German plan to phase out nuclear power and to create an energy system based on renewables is a strategy to achieve Sustainability by countering climate change but also by seeking economic growth, profits and the creation of new jobs. These arguments were important for generating public support for sustainable energies. However, this particular understanding of Sustainability was not initially shared by business makers.

Power supply companies have in many cases refused the idea of Sustainability for decades. They paid very little attention to renewables and rather continued their exclusive engagement in fossil and nuclear energy. For example, the chief technology officer of the German energy company E.ON, Klaus-Dieter Maubach, argued in 2011 that renewable energy would only very rarely be profitable in Germany and that his company would therefore rather invest abroad.[12] In 2012, the CEO of RWE, another big German energy producer even claimed that promoting solar energy in Germany would be as reasonable as planting pineapple in Alaska.[13] In 2015 both companies were in difficulties and will face massive additional costs when they dismantle their phased out nuclear reactors. In contrast, early innovations with regard to renewable energy were driven by grassroots initiatives and social movements in the 1970s and 1980s (Toke, 2011). Innovations often developed in niches and were undertaken by actors aiming to challenge incumbent energy regimes (Smith, 2012) Today, in contrast, sustainable energy practices and technologies are well positioned in the mainstream. In Germany, for instance, renewable energy now accounts for more than a quarter of gross electricity production.[14] Also the big energy companies mentioned above have turned towards renewable energies. E.ON established its own company (E.ON Climate & Renewables) in 2007 to develop renewable energy projects on an industrial scale and invested over 9 billion Euros in this field.[15] Similarly RWE founded RWE Innogy in 2008 with the explicit goal to 'vigorously grow renewable energies in Europe'.[16] Obviously, power companies have now taken up the understanding of Sustainability as an economic opportunity and consequently have translated the key idea into business models.

While initially sustainable energies were understood as a growing market with promising business opportunities, now, at least in Germany it is seen as essential for power companies to engage in Sustainability and '94% of CEOs in the energy industry believe that sustainability issues will be critical to the future success of their business' (Hanna & Lacy, 2011, p. 8).

The understanding of Sustainability as an opportunity for growth, with the side benefit of environmental protection is again based on the assumption of decoupling economic growth from energy use and its negative effects. On the one hand environmental impacts should be decoupled from energy production (mainly by renewables), while on the other hand energy consumption shall be decoupled from economic growth (mainly through energy efficiency). Finally a win–win situation is expected, to increase sales and profits, generate new jobs and allow social benefits, while environmental impacts are reduced. This understanding has been translated into multiple management practices, for instance the triple bottom line accounting (for an assessment see Norman & MacDonald, 2004). This concept was born in the 1990s and has gained some popularity in the sphere of business, investment and consulting. It advocates adding social and environmental aspects of a company's performance to the financial bottom-line.

Although the key idea of Sustainability was translated into multiple practices of business management, corporate governance or controlling, concerns remain that the idea of Sustainability is mainly mobilised by companies for legitimation or marketing purposes while not forcing them to address their ecological footprints or negative social consequences of their core business. This instrumental use of Sustainability to label one's economic activities as environmentally friendly or redefining Sustainability to suit one's practices are sometimes referred to as 'greenwashing'. From that perspective Sustainability can be depicted as entirely coherent with

> the continuation of today's energy industry, but also by means of the propagation of, for example, 'clean-coal' concepts to justify the building of new coal-fired power plants around the world. (…) In addition, nuclear energy receives the climate protection stamp, as if that makes all nuclear problems obsolete.
>
> *(Samuel, 2013, p. 64)*

These last examples give an impression of how the understanding of Sustainability as a business case gained importance and stability. Throughout this chapter the variety in translating the key idea of Sustainability into various new practices is evident. The chapter has elucidated the diversity, fragmentation, and contestedness of the concept of Sustainability, as well as the variety of its different manifestations in practice. Nevertheless, at least implicitly, patterns of stability (as outlined in Chapter 2) can be traced from a response to finite resources to a stronger emphasis on equity and development, and finally to the business case that increasingly takes up room. In the next chapter similar patterns will be investigated in modern food systems.

Notes

1 Strictly speaking, the sun is the essential source of most forms of renewable energy. Its energy can be used directly but is also the key factor for the existence of water cycles or the growth of plants that can then be used for energy production – plants that then can become fossilised over geological time-spans.

2 See http://www.agcs.allianz.com/insights/expert-risk-articles/energy-risks/ [accessed 29 September 2015].

3 See for example http://www.theguardian.com/environment/2012/feb/10/ grid-blackout-threat-renewables [accessed 29 September 2015].

4 Comprehensive discussions about environmental aspects of energy systems can be found in Dinçer and Zamfirescu (2012) or Vanek and Albright (2008).
5 See http://blogs.wsj.com/corporate-intelligence/2015/04/01/how-much-u-s-oil-and-gas-comes-from-fracking/ [accessed 29 September 2015].
6 See http://www.iea.org/aboutus/faqs/oil/ [accessed 29 September 2015].
7 For the USA see http://www.eia.gov/todayinenergy/detail.cfm?id=10191 [accessed 29 September 2015].
8 The SDC was an advisory body to the UK government, which existed between 2000 and spring of 2011.
9 See http://www.se4all.org/our-vision/ [accessed 29 September 2015].
10 See http://www.se4all.org/our-vision/ [accessed 29 September 2015].
11 See http://www.bmwi.de/EN/Topics/Energy/energy-transition.html [accessed 29 September 2015].
12 See http://www.spiegel.de/wirtschaft/unternehmen/e-on-vorstand-erneuerb are-energien-lohnen-sich-in-deutschland-nur-selten-a-764789.html [accessed 14 September 2015].
13 See http://www.taz.de/!5102707/ [accessed 14 September 2015].
14 See http://www.bmwi.de/EN/Topics/Energy/Renewable-Energy/renewa ble-energy-at-a-glance.html [accessed 29 September 2015].
15 See http://www.eon.com/en/about-us/structure/company-finder/e-dot-on-renewables.html [accessed 29 September 2015].
16 See http://www.rwe.com/web/cms/en/87202/rwe-innogy/about-rwe-inn ogy/ [accessed 29 September 2015].

5

SUSTAINABILITY AND FOOD SYSTEMS

Introduction

This chapter investigates modern food systems from a perspective of Sustainability. Food is among the most basic human needs. A massive range of human activities relate to the production, processing, and distribution of food. An equally broad and diverse range of practices relates to the preparation and consumption of food. Moreover, food is highly relevant with regard to Sustainability. For example, access to food (including land and water for food production) as well as the protection from food scarcity, low quality, or unequal distribution of food are age-old questions of social power and order. At the same time, agriculture (as the main source of food) always has to operate within ecological boundaries, which also implies that bad weather events, droughts, pests as well as overexploitation can have devastating effects on human health and well-being. Finally, contemporary globalised and industrialised food production as well as consumer cultures in affluent societies have created a broad range of additional challenges to Sustainability from processed food of low nutritional and health value to packaging, to massive amounts of food waste, and to so called food miles accumulated in production and distribution chains.

It is impossible to do full justice to all these themes and aspects in a single book, let alone a chapter of an introductory volume like this one. Hence, we are focusing our illustration on Sustainability with regard to agricultural food production. In contrast, we will say little about land, water, livestock, the processing of (convenience) food, packaging, distribution, retailing, and food consumption. Moreover, we have to gloss over many interesting details in order to sketch some major dynamics and constellations and to gain deeper understanding of the meanings and practices of Sustainability in this specific context, which is at the same time quite different as well as closely interconnected with climate change and energy. The chapter begins by sketching the formation and some of the key characteristics of modern food systems. Then, the key Sustainability problems are identified. On this basis, we select certain practices responding to these challenges. This discussion begins by tracing attempts of ecological farming and then moves on to struggles for justice and equity with regard to agricultural food production. The resulting image is a dual one: on the one hand, promoters of equity and justice have joined up with environmental activists and increasingly promote their values alongside each other. On the other hand, one can see the industrial extension of organic farming.

The formation of modern food systems

In order to map the Sustainability challenges of modern food systems the historic formation will be sketched very briefly as a background (for more detailed historical accounts, see Pilcher, 2006; Tauger, 2011). Human agricultural practices have always been strongly conditioned by the ecosystems surrounding them. Farmers mainly had to rely on natural resources available on-site, such as organic matter and farmyard manure to replenish soil nutrients. In addition, agricultural production was substantially based on manual labour. 'In these types of farming systems the link between agriculture and ecology was quite strong and signs of environmental degradation were seldom evident' (Altieri & Nicholis, 2005, p. 13). Farms were mostly independent, often small in size, and organised as family enterprises or cooperatives with strong local networks – if not controlled by large landowners or colonial powers. Seeds were mainly collected from own crops of locally adapted varieties. As a consequence the degree of food sovereignty and self-determination of farmers could be relatively high, if not suppressed by overriding social inequalities. At the same time, productivity and agricultural yields were determined and notably limited by ecological boundaries and natural conditions such as the availability of nutrients, the appearance of diseases or pests, as well as rainfall patterns and temperature. Hence, the productivity of agricultural practices was comparatively low, the risk of food scarcity high and often amplified by social inequalities.

The twentieth century has seen revolutionary changes in the field of agricultural production (for example, see Paarlberg & Paarlberg, 2000). The emergence of new agricultural practices on an industrial scale led to an explosion of efficiency and productivity. However, it also led to a whole range of new challenges to Sustainability, which will be addressed in the next section. Key developments originated in the utilisation of organic chemistry for agriculture that was particularly popularised by Justus von Liebig (Blondell-Mégrelis, 2007). It allowed the production of synthetic nitrogen fertilisers by the so called Haber-Bosch process that was commercialised by the German chemicals company BASF in 1914 (Smil, 2011). Based on chemical fertilisers, farming could overcome essential limitations in agricultural production, especially the availability of nitrogen for plant growth (Smil, 2004a). Soil restoration and crop rotation were not as necessary as before, which allowed for much greater specialisation and mono cropping. In addition the development of chemical pesticides and herbicides has played an important role to avoid pests and to control the growth of weeds. The early industrial evolution of food systems also involved growing farm sizes and relied upon the mechanisation and automatisation of agricultural practices. Productivity was further increased with the proliferation of new powerful machinery running on fossil energy as well as of more complex technical infrastructures, for instance to irrigate arid areas (Heinberg & Bomford, 2009). In the 1960s, this dynamic entered the next phase of the so-called 'Green Revolution'. This term originally referred to the breeding of new high-yield and more robust crop varieties, primarily of wheat and rice, suitable for intensive industrial farming (Hazell, 2003).

Based on the massive expansion of industrial practices the world agricultural food production increased remarkably; while food products became cheaper, the variety and abundance of available food grew, and famines could be reduced (Heinberg & Bomford, 2009). However, despite these advances, the industrial modernisation of agricultural food production has also raised questions and criticisms.

Dimensions of unsustainability

The benefits of modern food systems are obvious. However, modern food production is among those human activities creating the biggest environmental impacts (Heinberg & Bomford, 2009). Despite their productivity, practices of agricultural food production create multiple ecological as well as social concerns and are therefore highly ambivalent from the perspective of Sustainability. To begin with, agricultural food production contributes massively to the depletion of non-renewable resources, in particular, fossil energy needed to produce chemical fertilisers, pesticides, and to run heavy mechanical equipment and advanced technical infrastructures (see, for example, Woods et al., 2010). The energy return on energy investment (EROI, see also Chapter 4) illustrates how agricultural practices are dependent on external energy inputs. For instance in the USA it is estimated that '[a]pproximately 7.3 calories are used by the U.S. food system to deliver each calorie of food energy. Farming accounts for less than 20% of this expenditure, but still consumes more energy than it delivers' (Heinberg & Bomford, 2009, p. 2). Moreover, agriculture is responsible for about 70% of global water use (Millennium Ecosystem Assessment, 2005). Furthermore, food systems are crucial in terms of climate change as they contribute up to almost one third (figures range from 19% to 29%) of global GHG emissions, of which more than 80% are directly related to agricultural production (Vermeulen, Campbell & Ingram, 2012, p. 198). Soil degradation is yet another issue as the 'earth's soils are being washed away, rendered sterile or contaminated with toxic materials at a rate that cannot be sustained' (Oldeman, 1992, p. 19). Closely connected, the demand for additional arable land is a main cause for deforestation (Heinberg & Bomford, 2009). In addition to immense resource use, chemical intensive agricultural practices also contribute to increased pollution and a continuous loss of biodiversity. Agricultural chemicals also undermine food safety and quality, threatening the health of farmers as well as consumers (Vogt, 2007).

Moreover, the challenges to Sustainability in agricultural food production go beyond ecological problems or issues of food quality. First of all, inequalities in access to resources for food production (arable land, water, energy, etc.) and unequal distribution of food produce are a major cause for food scarcity and hunger. Despite remarkable increases in productivity, almost 800 million people worldwide still suffer from malnutrition and do not have access to enough food.[1] For example, the *Agriculture at a Crossroads: International Assessment of Agricultural Knowledge, Science and Technology for Development* (IAASTD), a review into the state of global agriculture with a particular focus on Sustainability, came to the conclusion that global food production had overtaken population growth but that the produced food was not available to those suffering from hunger, malnutrition, and poverty (IAASTD, 2009).[2] On the one hand, this development is interesting since availability of food was a major concern in *Limits to Growth* in the face of resource depletion and population growth. On the other hand, food systems are essentially social systems. Hence, a much greater range of factors than mere resource input influences outcomes. Regarding their social organisation and power, fewer but bigger producers increasingly dominate modern food systems (see below). This development is often criticised as undermining the economic basis of local farmers and rural communities. Every year millions of small farmers are pushed out of agricultural production, as they cannot compete with large-scale (often export oriented) producers – a trend with far reaching consequences (Heinberg & Bomford, 2009). Global corporations centralising and monopolising the provision of essential agricultural inputs (seeds, fertilisers, chemicals, machinery) and increasingly dominating the entire supply chain of food products (distribution and wholesale) raise concerns of food sovereignty and self-determination of farmers and consumers. Taken together, the multiple challenges related to modern food systems crucially impact 'the quality of life of the majority of the world's population and raise concerns about global food security in the long term' (Koc, 2010, p. 37).

This long list of challenges to Sustainability arising from contemporary agricultural food production is far from being exhaustive. However, it illustrates the diversity and complexity of interconnected problems. Moreover, it demonstrates the ambivalence of contemporary food systems, which are so much more productive than their pre-modern forerunners but also cause entirely new ecological and social problems – often of a much larger scale. Some of those concerns were already raised in the early stages of industrialisation, triggering various searches for alternative pathways long before the term Sustainability was defined (for the roots of Sustainability in agriculture see Hansen, 1996; Harwood, 1990). Moreover, these responses differ substantially with regard to their main criticisms, strategies, practices, and underlying values.

Ecological agriculture as caring for nature

Major alternative understandings of food systems and corresponding practices originated from ecological concerns and understandings of nature as fragile and in need of protection. It is beyond the scope of this chapter to provide a comprehensive overview of all related historical and contemporary understandings and practices.[3] Rather, we focus on a selection to illustrate how the different unsustainabilities of industrialised agricultural food production are contested, and how alternatives are sought.

A first example is *biodynamic agriculture*, which can be seen as the first organised movement in this field. A key founding moment were several lectures by Rudolf Steiner to a group of German farmers and land owners in 1924 (Vogt, 2007). Accordingly, the main motivations behind this movement are not purely ecological but also spiritual. Other authors extended Steiner's writings in the 1930s and 1940s and farmers started to put them into practice, mainly in Europe but also in the USA and Canada (Harwood, 1990). Today, the biodynamic movement is firmly institutionalised and a prominent player in the organic food market. In particular, its visibility is based on its 'Demeter' trademark in use since 1928 (Aschemann et al., 2007).

The next example is a type of farming that emerged in the 1940s, for which the term *organic* [4] was coined to describe an 'integrated, decen-tralized, chemical free agriculture' (Harwood, 1990, p. 8). For example, an early founding text, Sir Albert Howard's *An Agricultural Testament* (1943) strongly focuses on soil management and soil fertility, particularly addressing the processes of composting. Another English founding figure, Lady Eve Balfour initiated a long-term demonstration project on 'natural' farming in Suffolk, which started in 1939, ran for 25 years, and came to be known as the Haughley Experiment (see Balfour, 1976). It is also possible to identify institutions and infrastructures in this early phase, since the farming project was taken over by the UK Soil association only one year after its foundation in 1946. Today, the association claims to be 'the leading membership charity campaigning for healthy, humane and sustainable food, farming and land use.'[5] Moreover, while Howard and Balfour cited Steiner as an important source, their approaches were more explicitly scientific.

Organic farming was not limited to the UK but further institutions emerged in other countries. For example, the Australian Organic Farming and Gardening Society was founded in 1944 and the *Groupement d'agriculteurs biologiques de l'Ouest* in France in 1959. Moreover, the French organic farmers' association *Nature et Progrès* initiated the foundation of the International Federation of Organic Agriculture Movements (IFOAM) in 1972 (Paull, 2010).

Nevertheless, the overall share of organic farming remained at 'almost negligible levels until the 1980s' (Lockeretz, 2007b, p. 1). Before, it occupied a small anti-establishment niche (Geier, 2007). In addition, it was much more fiercely contested than today. For example, the US Secretary of Agriculture stated in 1971: 'before we go back to an organic agriculture in this country, somebody must decide which 50 million Americans we are going to let starve or go hungry' (quoted in Lockeretz, 2007b, p. 2). Only a few years later the US Department of Agriculture (USDA) published a widely recognised report on the limits and potential of organic farming, concluding 'that organic farming would receive an impetus from increasing concerns over energy shortages, declining soil productivity, soil erosion, chemical residues in foods and environmental contamination' (Lockeretz, 2007b, p. 2). Subsequently perceptions of organic farming shifted from eccentric to reasonable, as well as to being a legitimate theme for science and a relevant policy option that could gain public funding. The move from a niche project that was publicly bashed by policy makers to a major policy issue in food systems became particularly visible after an outbreak of the so-called mad cow disease BSE (Bovine spongiform encephalopathy) in the EU. The outbreak not only led to a massive breakdown in beef consumption, it also prompted the

German Federal Government to call for a massive shift in agricultural policy including a share of 20% organic farming – which has largely failed (though see Brand, 2011). At the international level, a global assessment report on agriculture initiated by the World Bank in cooperation with different international organisations also concluded that the extension of organic farming was necessary to reduce hunger and poverty and to 'facilitate equitable, environmentally, socially and economically sustainable development'.[6]

In the course of the growing acceptance and institutionalisation of organic farming, specific standards and labels were created to provide rules as to how organic products should be produced and to communicate these standards to consumers. Initially, labels were mostly administered by groups of producers, for example the Demeter label proves that a particular product was produced according to the rules of biodynamic farming. More recently, however, labelling of organic agricultural food products has even been taken up by governmental organisations. For example, the EU introduced a European label for organic food in 2000 and a new one in 2010; Germany followed in 2001, and the USA in 2002.

On the basis of the argument of this chapter so far, it has been demonstrated that agricultural food production is of immense relevance for Sustainability and that the industrialisation of agriculture has created a whole range of new challenges in this regard. Moreover, based on the examples described in this section, there have been different attempts to respond to these challenges by promoting agricultural practices that 'care' for nature rather than just exploit it. These responses differ in many aspects but overlap in others. For example, some have paid more attention to science, some to what they saw as traditional farming practices, some to spiritual ideas. These developments go back much further than the explicit discourse around the composite concept of Sustainability. Yet these perspectives agree on a clear ecological vision that was also expressed by Rachel Carson or Limits to Growth as described in Chapter 2. Not surprisingly, alternative movements explicitly referred to Sustainability very early to express their understandings and objectives. For instance, IFOAM's first conference 'was entitled "Towards a Sustainable Agriculture" – in 1977, long before the Brundtland Report and the Sustainability Summit in Rio de Janeiro in 1992' (IFOAM, 2012, p. 7). One of the first publications expressing the idea of

Sustainability in the context of food systems was *New Roots for Agriculture* (1980) by Wes Jackson (Kirschenmann, 2014). It took some time to combine these ecological concerns with the quest for social justice and global equity, which together form another main element of Sustainability. Yet, far from being absent, this element of Sustainability is very closely connected to agricultural food production. However, it was long carried by different actors, and ecological and social objectives often seemed to be in tension.

Agriculture, food, and the quest for equity

It has already been mentioned that issues of agricultural food production were always related to questions such as: Who has access to arable land and water? How is food distributed? Who is suffering most from food shortages? And also who profits from what type of agriculture and who does not? All of these questions essentially point towards issues of equity and justice. Moreover, similar to the ecological problems mentioned above, the industrialisation and globalisation of agricultural food production also created a whole set of new challenges to Sustainability. Several actors, concepts, and practices can be identified as trying to respond to these problems.

The *fair trade movement* serves well as a first illustration in this context. Contemporary ideas and practices have diverse historical roots. In the 1940s, US Mennonites started to trade products to support local producers in developing countries (Hockerts, 2005). This led to the foundation of Ten Thousand Villages in 1946, which is among the biggest international fair trade organisations today.[7] In Europe, the idea of fair trade was linked to the student movements of the 1960s and motivated by critiques of inequalities caused by global capitalism, free markets and international trade, systems that were assumed to be 'fundamentally flawed and … the only way to make them fairer was to set up a parallel (or alternative) trading model' (Redfern & Snedker, 2002, p. 5). The Oxford Committee for Famine Relief (Oxfam) was particularly important in pioneering fair trade in Europe. In the 1960s, the NGO systematically started to import products from the South to improve income opportunities of producers. It was followed by newly established fair trade organisations in other European countries, such as Germany and The Netherlands (Hockerts, 2005). While fair trade mainly began with handicrafts, soon an agricultural assortment was included, starting with coffee and tea and gradually expanding to integrate for instance chocolate, nuts or dried fruits (Redfern & Snedker, 2002). Today a remarkable range of fair trade (food and non-food) products is on offer. This expansion of the market share was crucially supported by the development of labelling and certification schemes. The first one was the Max Havelaar label, created in the Netherlands in 1988, and several certification schemes and labels followed (Redfern & Snedker, 2002). At that time fair trade was almost exclusively concerned with global justice, solidarity, and charity. Many organisations had Christian roots. Unsurprisingly, the equity centred notion of sustainable development as defined by the Brundtland Report in 1987 that partially originated in this milieu (see Chapter 2) was well received.

Concerns about fair trade were long seen as disparate if not in tension with ecological concerns. Meanwhile, however, fair trade organisations have taken up a comprehensive meaning of Sustainability that is close to that of the Brundtland Report and merged their social concerns with ecological objectives. For example, in its report *Fair Trades Contribution to a More Sustainable World*, Fairtrade International emphasised its commitment to ensure 'that the carrying capacity of ecosystems is not affected by agricultural production, as this would have a direct impact on producers' sustainable livelihoods' (Fairtrade International, 2010, p. 1). Similarly, Hockerts has claimed that 'although its primary concern was the improvement of social conditions among smallholders, fair trade nonetheless often increased the eco-efficiency of coffee production' (2005, p. 6). It is obvious that both the organic and the fair trade movement are challenging unsustainable practices in modern food systems. Whether or not their different understandings and emphasis can easily be aligned or are potentially conflicting, they are increasingly associated.

While the organic movement currently goes further in revealing the ecological conditions of production and the fair trade movement goes further in revealing the social conditions of production, there are signs that the two movements are forging a common ground in defining minimum social and environmental requirements.

(Raynolds, 2000, p. 297)

This common ground is expressed through organic and fair trade labels that are increasingly used side by side on certain products. Moreover, in some cases, the two movements also collaborate on an institutional level, for instance when campaigning for specific European policies.[8]

The *food sovereignty* movement constitutes another example of actors primarily addressing social challenges to Sustainability who are also linking their campaigns for justice, well-being, and self-determination of farmers with ecology in powerful ways. This movement took shape in the 1990s mainly supported by activists of Via Campesina, an international alliance of – among others – peasants, small-scale farmers, agricultural workers, rural women, landless people, and indigenous communities. A key text and an essential reference point in this context is the 'Declaration of Nyéléni' adopted in 2007 by the World Forum for Food Sovereignty held in Mali. The declaration gives a definition of food sovereignty that links equity and ecology with regard to food in a very tight manner:

> Food sovereignty is the right of peoples to healthy and culturally appropriate food produced through ecologically sound and sustainable methods, and their right to define their own food and agriculture systems. It puts those who produce, distribute and consume food at the heart of food systems and policies rather than the demands of markets and corporations.
>
> *('Declaration of Nyéléni', 2007)*

Moreover, food sovereignty is seen as requiring active political resistance, particularly, against corporate power, and should lead to 'new social relations free of oppression and inequality between men and women, peoples, racial groups, social classes and generations' ('Declaration of Nyéléni', 2007). This formulation also points to an important characteristic distinguishing the food sovereignty movement from the previous examples mentioned in this chapter. Promoters of organic farming have developed a wide set of concepts to distinguish their products from conventionally produced food, a broad range of practices how these organic products can be produced and distributed, as well as a system of practices of labelling and certification that are more technocratic but closely related to Sustainability. The situation is similar with regard to fair trade, which is about establishing alternative trade networks in theory and practice, as well as about ensuring that these networks correspond to the underlying values und ideas. Certificates, standards, and labels are instruments of particular importance in this context, too. Food sovereignty, however, is much more about political networking, mobilisation, action, and resistance. It is an explicitly political project that tackles power asymmetries.

Overall, the adoption of ecological values by justice movements, in particular from the global South, is less remarkable as an intellectual development. This can be traced back before the Brundtland Commission. Rather, it is interesting to see how the same political compromise is again taken up 30 years after Brundtland and in a context where grassroots actors from the global South are so prominent. Just as a reminder, Chapter 2 also described how ecology was still perceived as a threat to development by many governments, for example, during the Earth summit in Rio 1992. Moreover, it shows that the comprehensive framing of Sustainability as equity and ecology also creates intellectual and political resources to define an alternative world in which food sovereignty could be realised. Thus, this notion of Sustainability also implies suggestions as to which alliances could be formed in order to fight for such an alternative world of just and ecological food production and distribution.

In fact, some traditional environmental organisations have moved beyond a narrow framing on ecological Sustainability focusing on resource depletion and environmental degradation. Instead, they have started to pursue their objectives in a context of social justice and global equity. For example, the website of Friends of the Earth International does not list a single environmental issue without putting it in the context of broader social and political inequality, as well as naming specific power asymmetries that need to be resisted.

While the example of food sovereignty illustrates a very political response to meeting Sustainability problems, the practices of organic farming and fair trade are also closely tied to the market. In fact, success in the market is an important sign for success of organic and fair trade products, such as coffee. The reason might not only be the insight that environmental standards in production are crucial for the livelihoods of producers but also that organic certified products allow a higher price and can serve as a marketing strategy to increase sales. This perspective points to another dynamic parallel to the political struggles for food sovereignty. This refers to the promotion of Sustainability in the market by creating sustainable products and so called ethical consumption. In short, Sustainability is not only about ecological limits and global justice but can also be understood as an economic feature to establish a business case and achieve a kind of green growth.

The agricultural green economy

Food is not only an essential human need but also an increasingly important aspect of different lifestyles, which are often tied to specific products. In contemporary consumer societies, food gained importance as an expression of values. Practices of buying, preparing, and eating (sometimes also growing) food have become essential expressions of certain ways of living. Some consumers seek lifestyles that should put Sustainability in practice (for sure, many others do not). They include, for instance, people living 'Lifestyles of Health and Sustainability', so called LOHAS (see Emerich, 2011), the slow food movement, and people adopting vegetarian and vegan diets for ecological as well as animal welfare reasons.

Despite the diverse activities undertaken to tackle the aforementioned ecological and social impacts of industrialised food systems, the dominant political and cultural climate favours activities compatible with markets. 'From the perspective of green consumerism, the organics market is a powerful engine for positive change because it promotes greater environmental awareness and responsibility among producers and consumers alike' (Allen & Kovach, 2000, p. 221). From such a perspective, consumers are not only assumed to be interested in healthy and safe food, but also are seen as crucial if not responsible for realising Sustainability by demanding and consuming sustainable products in shops and restaurants.

In fact, demand for sustainable food products has grown substantially in recent decades. (The following review of market development draws on Aschemann et al., 2007.) This development was rooted in the environmental movements and benefited from an increasing awareness of environmental and health aspects with regard to food products. The standardisation of organic farming practices and products as well as newly established certification and labelling schemes provided not only information for consumers but also made a whole sector legible and accessible to market actors and regulators. Moreover, subsidy programmes and information campaigns for organic farming increased institutionalisation of production and distribution networks, and various food scandals contributed to growing awareness of organic food. More consumers were motivated to pay premium prices for products of high quality and safety. The significant growth of the market for organic food products finally convinced mainstream retailers to offer organic products, which definitely helped to further increase the presence of organic food in the food system as a whole. As a result today's consumer demand for sustainable food products has become a relevant economic factor.

The promising growth prospects of sustainably produced food products in combination with increased political support for this niche also motivated agro-industrial companies and large retailers to enter the organic food sector. In this case, markets for organic food are served by highly specialised food production on an industrial scale. While the farming practices of sustainable yet industrial agricultural producers might still rely on heavy machinery, mono-cropping, and high resource input, including extensive packaging and food miles, they largely reduce pesticide use thereby reducing their environmental impact and health risks.

In this manner, the EU Commission adopted an Action Plan for Organic Food Production, which defines the production of organic food as

an overall system of farm management and food production that combines best environmental practices, a high level of biodiversity, the preservation of natural resources and the application of high animal welfare standards (…) in line with the preference of certain consumers for products using natural substances and processes.

(European Commission, 2014a, p. 2)

The plan further emphasises the need to support the growth of the organic food sector, which will promote the dual objective of environmental Sustainability and economic growth. What is remarkable about this development is that organic food is being redefined from a radical niche to a promising growth industry.

On this basis, even agro-industrial giants such as Monsanto can now claim to promote Sustainability through their own agrifood practices and products. For example, a recent sustainability report published on the company's website declares that

> Sustainable agriculture is at the core of Monsanto. We are committed to developing the technologies that enable farmers to produce more crops while conserving more of the natural resources that are essential to their success. Producing more. Conserving more. Improving lives. That's sustainable agriculture.
>
> *(Monsanto, 2015)*

Today the idea of Sustainability is also mobilised by industrial food producers supplying organic food on a large scale as well as by international corporations seeking to further increase the produced crops by new technologies including genetically modified organisms (GMOs) and more effective pesticides and herbicides. At the same time fair trade organisations refer to the idea of Sustainability in the marketing of their products, and large trade chains are offering organic product lines at competitive prices. The resulting image of the meanings and practices relating to Sustainability in the particular context of agricultural food production has not become simpler but rather has gained another layer of diversity. Most generally speaking, Sustainability in this context is about sustaining food systems, which need to provide safe, healthy, and enough food for a growing population. Having a closer look into the field of food production and consumption reveals that the idea of Sustainability is highly fragmented and originates from multiple problem understandings and normative prescriptions. Science is approaching and analysing the challenges of food systems in different traditions and from multiple perspectives. At the same time there is little agreement on appropriate political responses (see Lang & Barling, 2013). As the multiple responses to the concerns of modern food systems illustrate, different actors understand Sustainability in very different terms. Therefore, the next and at the same time final chapter of this book moves back from specific challenges to Sustainability in order to discuss the nature and relevance of this key idea. In particular, it tries to make sense of the essential diversity and contestedness of Sustainability in theory and practice.

Notes

1 See https://www.wfp.org/hunger [accessed 5 October 2015].
2 See also https://www.wfp.org/hunger/causes [accessed 5 October 2015].
3 For the history of alternative movements in agriculture, see Harwood (1990) or Kirschenmann (2014). For an international history of organic agriculture see Lockeretz (2007a).
4 The term 'organic farming' is said to be coined by Lord James Northbourne (1940) who also played an important role in bringing Steiner's ideas about biodynamic farming to the UK (Paull, 2014).
5 Taken from http://www.soilassociation.org/aboutus/whoweare [accessed 5 August 2015].
6 See, http://www.unep.org/dewa/Assessments/Ecosystems/IAASTD/tabid/105853/Default.aspx [accessed 20 December 2015].
7 See http://www.tenthousandvillages.com/about-us/ [accessed 20 December 2015].
8 See http://www.ifoam.bio/en/news/2015/04/23/ifoam-organics-interna tional-advocating-fair-eu-organic-import [accessed 5 October 2015].

6

SUSTAINABILITY AS TRANSFORMATION AND REFLEXIVITY

After the more detailed investigations of challenges to Sustainability arising in the context of climate change, energy, and agricultural food production, this final chapter moves back to the more general level of analysis of the first two chapters. Its aim is to reflect again from such a more comprehensive view the role that the idea of Sustainability could ideally play in the contemporary world. For this purpose, this chapter will first draw on the previous three chapters in order to identify broader features and patterns characterising Sustainability and its historical development. Then, it will discuss the nature of Sustainability as an idea and how it can contribute to a more sustainable order.

Making sense of the essential diversity of sustainability

To begin with, the historical overview and the subsequent more detailed illustrations focusing on particular challenges to Sustainability demonstrated that – below the most general meaning as the aim to seek a balanced relationship between human striving for well-being and prosperity with its ecological, and socio-cultural implications – diversity and contestedness are essential characteristics of Sustainability. Therefore, Sustainability can occupy a rather large range of different meanings and practices across time and in different contexts. This characteristic does not have to damage the power and relevance of Sustainability. However, it needs to be acknowledged and understood to navigate the essential plurality underlying the idea of Sustainability.

The theoretical view on Sustainability as a social phenomenon constituted by practices offers important insights for this purpose. For example, it can explain why the notion of Sustainability is so stubbornly resisting against being defined in more specific and stable terms. Instead, a plurality of definitions and emphases always (re-)emerges. Relatively closely knit constellations of actors that form around a particular set of problems can be quite stable over time.

For example, in Chapter 3 it was found with regard to climate change that the particular notions of a green economy, green growth, and technology-based ecological modernisation developed into a stable pattern that was described as technological voluntarism. This optimistic project seeks transformation on the basis of free market instruments and the continuation of societies focused on an expansion of income, products and services. This dominant pattern emerged despite the objection that there still is no solution to the challenge of how to absolutely decouple economic growth from ecological impact. The stability of the green economy idea in the context of contemporary discourses on Sustainability is less grounded in its actual feasibility but rather in its stable connection to other elements in the Sustainability debate, such as the notion of inclusive economic development through more justly distributed economic growth. With regard to agricultural food production, Chapter 5 has illustrated how difficult it was to merge different understandings of Sustainability. For example, although the fair trade and the organic farming movement are both addressing dimensions of unsustainability in food production, their understandings are emerging from and translating into very different practices. The very fact that meanings, ideas and strategies are always embedded and reproduced in social practices can cause inertia, as complex and long-standing constellations of practices may need to undergo transformation in order to allow a recombination of different ideas of Sustainability. However, despite the relative stability of single projects (such as organic farming) and their related practices, new meanings can always be born out of changes in the associated practices. At the same time, a shift in the understandings that are an essential element of established practices can allow or even induce a change of certain routines. This is what happens, for instance, when activists of the fair trade and organic farming movements undergo attempts to transform their practices in a way that reflects both understandings and the respective strategies. The result might be an entirely new and more inclusive understanding of Sustainability.

Needless to say, all of the more or less stable practice constellations outlined in the three chapters above have to interact with understandings, interests, and practices of other actors, which might create tensions, conflict, and the need to find allies and compromises. Moreover, the challenges to Sustainability can be very dynamic and change quickly, for example, due to new evidence about the underlying problems of the pursued strategies, or due to unexpected developments such as (natural) disasters, financial crises, or political developments that change the institutional conditions underlying how Sustainability can be promoted.

On a slightly broader scale, Chapter 2 has shown how the idea of Sustainability stabilised around particular constellations of meanings, challenges, strategies, and instruments. In part, this is due to the fact that even the most basic ideas of social thought do not exist outside social practice. Rather, they are defined, explained, negotiated, contested, and refined in discursive and literary practices such as the writing of policy papers. For example, this can be observed when the Intergovernmental Panel on Climate Change (IPCC) defined two degrees of global warming as the crucial limit that should not be exceeded or when the United Nations (UN) installed the World Commission on Environment and Development (WCED) and asked it to write a report. A view on social practice helps us to connect definitions and analyses with the actual practices of writing, giving speeches, holding conferences. Such a view on how meanings of Sustainability are used in practice also allows for linking them to the respective agents, their worldviews, interests, and position in terms of power and location. For this insight we can draw on a rich body of literature in intellectual history of ideas (for example, the so called Cambridge School) as well as in social studies of science, technology, and society. These perspectives have emphasised the need to investigate how knowledge is being made, circulated, and received in practice, how it is carried by social agents, how evidence is mobilised, and how it is defended and stabilised in specific social contexts. Ludwik

Fleck, an early ancestor of science studies described those social contexts in which knowledge is developed, shared, and debated as 'thought collectives (*Denkkollektive*)' (Fleck, 1980, p. 54). Later, more recent accounts of the stabilisation of knowledge through the stabilisation of social and material orders and *vice versa* was described in terms of actor-networks (Latour, 1987, 2005), translation (Callon, 1986), epistemic cultures (Knorr Cetina, 1999), and co-production (Jasanoff, 2004). Jasanoff's work is particularly instructive in this context. She argues that social order and knowledge are always produced alongside each other and constitutive of each other. Against this background, it is of no surprise if different groups of people share different versions of a particular idea. If they have to find compromises with others they have to consist of both a knowledge dimension and a political dimension.

Sustainability is a social phenomenon that is also carried, mobilised, refined, contested, and defended by different thought collectives originating in different cultural contexts, for example, climate scientists, energy producers, or food sovereignty activists. All of them bring very different experiences, interests, norms, values, and power with them. Therefore, rather than seeking the essence or the objective core of Sustainability we need to acknowledge that the idea of Sustainability is just the sum of its manifold uses in practice and that there might be significant difference across time and different contexts. However, Sustainability is not entirely open and arbitrary. Its coherence is based on what Ludwig Wittgenstein called a family resemblance (Wittgenstein, 1953) rather than an essential core.

In addition to the diversity of groups and agents engaging with Sustainability, its meanings are also affected by the different practices pursued in particular instances. In this context, important practices are, for example, identifying Sustainability challenges, and institutions, strategies and instruments to tackle them. Moreover, they can include the setting up of indicators, monitoring systems, and feedback processes to evaluate the effectiveness and consequences of such instruments. In addition, practices of resistance are an important part of these practices since design, the values inscribed, and assessments of the consequences of the practices mentioned in the previous sentences can differ considerably and become quite contested. The link between these discursive practices and attempts to put them into reality can be very complex. For example, it is one thing to formulate political goals for renewable energy in the EU, to commission research reports and computer models how such an energy system could look like. However, it is quite another thing to put such a system into place, to extend it, and to maintain it; this requires that a plethora of different practices from contexts as diverse as engineering, finance, energy accounting, network regulation, and consumption are re-aligned and invented if necessary.

Finally, the relationship between the idea of Sustainability and its elements and respective measures to put it into practice is not a unilinear causal arrow from concept to action. In contrast, Sustainability might be read out of practices that are not explicitly linked to it.

Sustainability as epistemic commons and experimental transformation

It should be clear by now that Sustainability is so much more complex than a clear-cut rulebook that could simply be 'applied'. So, given this essential characteristic of Sustainability, what relevance does the idea have and what role could it play in contemporary societies struggling not only with climate change but also with their ecological impacts more generally, as well as with the unequal distribution of life-chances and well-being? We claim that it is more appropriate to think of Sustainability not as a single entity but rather as a specific space for debate, analysis, and action that can – and should – be an important source for social reflexivity. The discursive practices and institutions which conceptually elaborate Sustainability, for example, constitute this space by academic and political discourses, indicators and monitoring systems, as well as by management and consumer practices aiming for greater sustainability. Moreover, material elements, for example, soil, forests, animals, and oceans, as well as waste and GHGs, are also part of this arena. An abstract concept like Sustainability does not hold without those material elements that are absolutely necessary to identify and assess challenges, to devise remedies, and to evaluate possible progress or degradation. For example, it is worth remembering that *Limits to Growth* was mainly an analysis about the material impact of human life on the earth's ecosystems and, thereby, on the living conditions of humanity itself. The question whether and how prosperity and well-being could be possible without ever increasing material impacts in terms of environmental degradation, resource depletion, and emissions is a key question triggered by the debate about Sustainability.

Climate change might be a good case for Sustainability as epistemic commons. As stated in Chapter 3, climate change as an issue left the scientific realm transgressing into politics and economics. Moreover, the global attempts to address climate change also led to the establishment of a new and interdisciplinary global epistemic community. The IPCC plays a central role in this development but also provides links with different parts of societies beyond science. Moreover, the hybrid nature of this institution at the science–policy–society interface also demonstrates the normative and contested nature of Sustainability as a political project for the twenty-first century.

From a more general perspective, a central value of Sustainability is that it prompts questions and frameworks for assessing them while clear-cut solutions have to be developed in the context of specific challenges, institutions, and practices. In fact, this is the central argument of this concluding chapter: the idea of Sustainability is an important reference point that can help to reflect on 'our' relationships to the environment, the economy, and those people in greatest need. Moreover, the space created by the multiple and diverse practices is the space where problems and questions of Sustainability are identified, scrutinised, tackled, and where different solutions are evaluated with regard to their appropriateness and effectiveness. There will always be actors referring to Sustainability for instrumental reasons (Stirling, 2009), for example, 'green washing' accounts of their own activities. However, the history of Sustainability offers enough powerful resources to counter such strategic misuses of Sustainability. Promoters of ecology and social justice have found themselves at the weaker end of asymmetric power relations most of the time – but nevertheless have developed concepts and practices to promote this idea and to keep it alive. From this perspective, the contestedness of Sustainability is not a weakness but a strength.

Moreover, similar to other key ideas, such as democracy or justice, this view on Sustainability implies that it is a public issue rather than a private concern. Accordingly, the normative and analytical knowledge developed with regard to this idea are in fact public goods or, in other word *epistemic commons* (Morisse-Schilbach & Halfmann, 2012); they can only be achieved collectively. Lively debates and practical activities analysing specific challenges, identifying possible solutions, and aiming to make existing practices in different contexts more sustainable, as well as resistance and critique of existing problems of unsustainability, can actually be seen as improving the substance of Sustainability. For example, contestations can make vested interests and power asymmetries visible or reports by affected citizens can contribute to much deeper understanding of particular Sustainability challenges. This can be seen as a form of collective intelligence to be nurtured by society – similarly to the role of democracy in pluralistic and diverse societies. However, while democracy might be content with checking power and enabling a degree of self-governance in the present, Sustainability provokes questions about the future. This anticipatory potential could even help a society to formulate its expectations about possible and desirable futures and seek transformative practices to pursue them. In fact, there is a growing and increasingly institutionalised repertoire of practices aiming at the empowerment and the participation of citizens. In part, they are meant to foster accountability and legitimacy of science. In part, they are meant to provide an element of collective intelligence. They get the opportunity to debate policy or technological options, pathways, and visions. Methodologies of this kind include, for example, Q-Methodology (Barry & Proops, 1999), deliberative exercises (Felt & Fochler, 2010; Irwin, 2006; Pallett & Chilvers, 2013), and constructive technology assessment (Rip, Misa, & Schot, 1995). This tendency towards increased participation is due to citizen's critical demand but can also be justified by the argument that increased participation and a pluralisation of viewpoints can increase the substantive quality of decision making and its legitimacy.

Individualistic views on Sustainability, for example, views on a world where ethical consumers are made responsible for realising Sustainability by making individual decisions in the market, will always be of limited scope. However, public goods need to be produced, maintained, and reproduced in practice – and the same is true for the idea of Sustainability. This view on Sustainability as epistemic commons implies a very dynamic conception of Sustainability. It is less about reaching a stable equilibrium but more about constant scrutiny, analysis, and reflection of challenges, goals and success.

Transformation, understood as a collective search for a new order and new relationships (between people but also including technology and ecosystems), is hardly ever a planned process where direction, speed, and outcome are under full control. In contrast, it is an open-ended collective search for moving targets. Most of the meanings and practices related to Sustainability are concerned with present problems, which require some form of transformation – for example, political, economic, and technical as well as in the sphere of citizens' everyday life – to move towards a different trajectory. Putting Sustainability into practice is far more than simply implementing objectives or strategies. Rather, it is about experimenting with practices that may elaborate Sustainability conceptually, reduce the negative effects of human activity in specific contexts such as energy or food production, and aim for monitoring and adjusting the development of these experiments. For example, mitigating and/or adapting to climate change, building more sustainable energy systems based on renewable resources, or finding ways to produce food in ways that ensure food safety, fair distribution, and minimising ecological impacts have to be seen as open-ended searches.

There is a long debate on experimentation as a mode of public policy and management, which cannot be presented in detail here. The open-endedness and contestedness of most attempts to realise Sustainability imply that it is impossible to assess the outcomes and risks of those attempts with full certainty in advance. Therefore, reflexive governance (see, for example, Ely et al., 2013; Stirling, 2009; Voss, Bauknecht, & Kemp, 2006), which is geared towards such experimental processes regularly has to check for unwanted side-effects of chosen activities, whether they still lead towards the envisioned goals, and whether alternative activities or goals should be chosen. Often, reflexive governance just has to 'open-up' dominant power relations (Stirling, 2008).

Infrastructures for Sustainability

The diverse understandings and practices of Sustainability create an important space for collective imagination and experimental transformation. As such, Sustainability could crucially contribute to social reflexivity and allow a (local, national, global) community to envisage and debate its options and responsibility in the present and future. However, we have also seen that Sustainability is fragile and not of utmost priority in many contexts. Therefore, in order to fulfil its reflexive potential, certain conditions should be in place.

Furthermore, the previous chapters have shown that Sustainability is not without competitors and that organic farming, renewable energy, and even the – meanwhile well-grounded – conceptualisation of anthropogenic climate change have faced significant opposition. Moreover, the experimental and reflexive potential of Sustainability can suffer if the latter is being defined too narrowly in either technocratic or consumerist terms or when central challenges are not identified. Therefore, Sustainability requires certain infrastructures in order to actually do its work as frame of reference within a particular community. The diversity of meanings and practices demonstrated in the previous chapters also implies that Sustainability takes different forms in different places. On closer inspection of particular contexts and perceived challenges, different infrastructures can be found. But before we give examples from the previous chapters, we assert that there are several types of infrastructures. Many of them do not correspond to the conventional image of large-scale technological hardware such as streets and electricity lines.

First and probably most important from a historical point of view, civil society organisations have been of extraordinary influence, in particular, regarding ecological issues but also regarding activities to fight global poverty and promote equity. Organisations like Greenpeace and Friends of the Earth were key actors for the emerging environmental movement and have carried this cause through the years. Some of these organisations have turned into global players influencing understandings of Sustainability on a global scale. Some are active at the grassroots level promoting Sustainability practices in North and South. Today, the website of the UN Department of Economic and Social Affairs lists 9,169 civil society organisations concerned with sustainable development, 3,939 of them NGOs.[1] This does not mean that social movements and NGOs are always right. Civil society is diverse, spans a broad political spectrum, and also contains agents who oppose Sustainability, for example, by denying climate change. However, civil society organisations concerned with ecology and/or equity are crucial agents with regard to the particular idea of Sustainability, its (relative) historical success and especially for the practices that have evolved in relation to it.

As a second example of this kind, Sustainability is closely tied up with certain documents that have become stabilised representations of Sustainability at large. Among the most important infrastructures of this kind is the report *Our Common Future* by the Brundtland Commission. It became such an important point of reference that it still provides one of the best-known definitions of Sustainability. The two-degree limit set by the IPCC has gained such importance that it can also be seen as an infrastructure behind stabilising and ordering meanings and practices related to Sustainability. These representations are infrastructures because they enable and inform action. This does not mean that they have causal powers but that many practices – explicitly or implicitly – draw on the meanings provided by them. Moreover, they are available as important reference points when particular understandings, practices, and consequences of the latter are assessed or contested. For example, resistance to unsustainability is much easier if it can be legitimated by assessments of institutions such as the WCED or the IPCC.

Therefore, institutions and organisations constitute a third important type of infrastructure. Again, there can be different kinds of organisations and institutions that can play important roles as infrastructures for different meanings and practices of Sustainability. For example, they can be state institutions at national or regional level – government branches, parliamentary committees, (mainly green) political parties, regulatory agencies, or advisory bodies such as the Sustainability councils and commissions that exist in many countries. Moreover, the previous chapters have pointed to several institutions in transnational politics that contribute to monitoring, elaborating, and promoting Sustainability. The IPCC certainly is a particularly prominent example of this kind but most other multilateral environmental agreements also establish advisory bodies mediating between politics and science.

Fourth, science itself has seen the development of important infrastructures. The previous chapters have shown that Sustainability is very often framed in scientific terms and that its promotion at least to some extent needs to be grounded in scientific analysis and expert knowledge. And in fact, alongside the history of Sustainability one can detect the emergence of different disciplines, research institutes, and programmes concerned with these questions. For example, it was argued in Chapter 2 that the success of *Limits to Growth* was also due to its scientific nature. Today, climate change, sustainable energy transitions, or sustainable food production are all academic growth industries benefiting the natural, engineering, and social sciences. While scientific experts are important to deal with the complexity of the most pressing challenges to Sustainability, the same complexity also implies that moral, political, and natural-scientific aspects are so closely interwoven that science is regularly challenged to be more open and cooperating with societal actors. Furthermore, closely connected to science, education is also seen as a crucial tool, especially with regard to educating citizens to be aware and committed to Sustainability (see, for example, Barry, 2006; Dobson, 2007).

A final type of infrastructure concerns more specific instruments and procedures. This involves indicator systems, standards and labels to account for Sustainability – mostly based on quantitative measurements (Bowker & Leigh Star, 2000; on the importance of quantification, see Miller, 2005; see also Porter, 1995). At the same time, those kinds of infrastructures can also be more complex legal and policy products such as the EU Emissions Trading System (EU ETS) put in place to reduce industrial GHG emissions (see, for example, Voss & Simons, 2014). The German feed-in tariff to promote renewable energy production by private households is another example of this kind (see Hoppmann, Huenteler, & Girod, 2014). Different participatory techniques that were also mentioned above have been developed to enable the systematic participation of citizens in the experimental governance of sustainable transformations. Participatory practices are much more open and less easily disciplined than metric indicator systems or labels. At the same time these instruments are different from general debate in democratic institutions, the media, and public since they are based on standardised practices and understandings and help to produce accountable and legitimate outcomes. In other words, they are technologies of governance rather than free exchanges among equal citizens (see Felt & Fochler, 2010; Felt et al., 2012).

On the whole, such standardised instruments can travel and can be taken up by agents in the course of a learning process. This also requires not only people who integrate them into their everyday practices but also a certain flexibility on the side of the instruments so that they can be translated and adapted to work and to make sense in another context. Finally, instruments are also important because they can make Sustainability legible and governable – not only in bureaucratic and management contexts but also with regard to the wider public debate. This function is, for example, essential for assessing whether a particular strategy is an effective way to achieve a goal. At the same time, however, they are not neutral. Each instrument favours certain perspectives over others and is carried by specific constituencies, while other actors might have no or only limited access (see Voss & Simons, 2014).

Caring for Sustainability, caring for transformation

This section develops the second main argument of this chapter and the key take-home message of the book as a whole: whoever wants to promote Sustainability and realise it in practice has to actively care for this cause investing time, energy, and passion.[2] There are innumerable forms and possibilities how this care could be exercised. In fact, this book has gone to great lengths to demonstrate that Sustainability is a diverse phenomenon comprising multiple meanings, challenges, and activities. The diverse and contested nature of Sustainability implies that – as an idea – it is far from self-evident and does not exercise causal force. In contrast, those aiming to further Sustainability need to formulate refined meanings that apply to specific challenges and contexts. Moreover, as we have seen, the link between discursive practices analysing and defining Sustainability and acting on its behalf is not a direct one, it is equally important that Sustainability is built into the everyday practices in contexts where Sustainability problems need to be challenged. This requires actual work, time, and energy. Moreover, leading by example becomes a crucial aspect of all attempts to realise Sustainability in practice.

For example, caring for Sustainability in the context of climate change mitigation involves the production of scientific evidence about the global climate, the risks from increasing human emissions, as well as suggestions as to how this problem might be tackled. It further requires the negotiation of immensely complex political settlements, such as the Kyoto protocol, and of large-scale instruments to implement these agreements, such as the different national CO_2 accounting systems or the EU ETS. Finally, climate change mitigation requires that individual actors translate and integrate this objective into their everyday practices in bureaucracies, businesses, and households. Moreover, it will meet resistance and contestations, conflicts will have to be endured, alliances will need to be built, and compromises to be made. Even though those who categorically deny that climate change is caused by human activity have less and less influence – at least in the academic arena – climate change still offers more than enough issues of disagreement and resistance. For example, state governments struggle to agree about reduction targets, their legal enforcement, and on the distribution of responsibilities to cut emissions between industrial and post-colonial states. Scientists struggle how to deal with the hockey stick, or whether and how an absolute decoupling of environmental impacts from economic growth might be feasible. Integrating Sustainability into one's personal everyday practices can be very difficult because of one's embeddedness in several systems of practice and material infrastructures that may offer opportunities but also impose costs and resistances, for example, when public transport or organic food are not available or too expensive (Shove, Watson, & Spurling, 2015).

The diversity of understandings and doings with regard to Sustainability implies regular choices for specific meanings and practices with regard to Sustainability. However, since the diversity and contestedness of Sustainability are essential characteristics that cannot be reduced into a single definition nor resolved by reference to evidence, such choices are essentially political decisions. For example, these decisions also affect questions about who is in power, which challenges to Sustainability are most pressing, who has the power to define them, who is able to intervene in unsustainable practices (see also Smith & Stirling, 2010). Political situations require democratic participatory solutions. Therefore, engaging in democratic debate about the meanings and practices of Sustainability as well as of related challenges is the most basic and most important form of care for those aiming to see this idea being realised. The collective intelligence and reflexivity that might emanate from the interaction of different agents, meanings, and practices in the Sustainability arena is strongly based on a political process, and power might be as important as knowledge. Therefore, scrutinising whether certain positions, arguments, and activities in this context might be merely instrumental in order to pursue other interests is another aspect of why taking care for Sustainability is so important.

This chapter ends with a final important aspect of caring for Sustainability: positioning. In order not to get lost among the different meanings and practices of Sustainability, and in order to be able to act in this social as well as political arena, it is important to reflect one's position and to communicate it to others (drawing on, for example, Haraway, 1988). On this basis, certain aspects of Sustainability can become associated with certain agents who choose to represent and take responsibility for them; for example, activists who engage in climate camps and climbing shovel excavators to hinder coal extraction or who block trains to cut power plants off from their 'dirty fuels', as Greenpeace did in the UK (see Smith, 2012), in order to destabilize incumbent regimes from a relatively weak power position. Another example is energy cooperatives, founded and run by local groups of citizens, investing, setting up and running wind farms and biomass power stations to produce their own energy, aiming to realise a more sustainable energy future in practice. Final examples from the context of food and agriculture might be farmers who risk growing old varieties and experiment with long forgotten agricultural practices for reasons of Sustainability, or local groups who establish fair trade shops and markets to provide the producers in the global South a better chance for just incomes and livelihoods. Moreover, alliances can be formed and more specific sustainability projects such as organic farming or renewable energy can be promoted.

This book is an attempt to give an account of Sustainability that is as comprehensive and as brief as possible at the same time. We end it by outlining our own position within the arena of understandings and practices of Sustainability. This is to remind readers that our view on Sustainability is a view from somewhere and by someone. It should encourage them to reflect and to find their own position.

We begin by stating that we are rather sceptical about the possibility of decoupling economic growth from growing resource use and environmental degradation. In the same way we are highly sceptical about the potentials of high technological fixes that are often promoted on the imagination that we just need the right machinery in order to solve most sustainability challenges while basically pursuing the same lifestyles as today. In particular, this scepticism applies to nuclear power, climate engineering, and GMOs. It is, first, based on the uncertainties related to these technological responses to specific sustainability challenges. This involves unwanted consequences like nuclear waste or accidents. Equally important, the applications and meanings attached to these technologies are far from limited to remedying unsustainability (see Hecht, 2009; Jasanoff & Kim, 2009). In contrast, for example, all of them could also be explicitly used to cause harm. Finally, the risks involved have a global reach, which makes democratic control and choice impossible as well as experimental scrutiny difficult. Once in place, these technologies leave very little space for autonomy and self-governance, especially, for those who do not want to take these risks. As a final sceptical remark, we would claim that individualistic strategies and instruments are less capable of promoting and realising Sustainability – especially, if they are primarily imagining society as a market place and citizens as consumers.

In contrast, we think it is essential that caring for Sustainability imply caring for public debate and democratic governance. Practice and meaning are social and inter-subjective phenomena. In fact, the whole book has been devoted to investigating Sustainability as an idea, a social phenomenon. Therefore, any attempt to promote and to realise Sustainability needs to imagine it as a collective and public endeavour and to accept its political nature if it does not want to be incomplete. This political nature implies that caring for Sustainability will often have to take the form of political activism rather than consensus seeking or calm problem solving.

Furthermore, since we still do not see how economic growth could be decoupled from growing resource use, emissions, and waste, we hold that attempts to be more sustainable should aim for sufficiency – in lax terms doing less with less things – rather than business as usual of a greener kind. In line with the Brundtland Report, we even believe that, in particular for Western consumerist societies, sufficiency could be more than an unavoidable means to an end. Doing less with less stuff could instead imply higher quality of living, less stress, and greater self-determination. We are going so far as to expect a possible double dividend from aiming for greater sufficiency, namely the possibility of living better while being more sustainable (see T. Jackson, 2005).

We will close this discussion by reminding our readers that the position of its authors needs to be fed into the interactive 'mangle of practice' (Pickering, 1995) or, in other words, into the contested and diverse debates and practical attempts to realise the idea of Sustainability. More generally, Sustainability is a field of meanings and practices related to envisioning better worlds and actions devised to realise them. This does not mean that these visions will ever be realised exactly as imagined. In the worst case, they might be lost among different projects struggling over the right meanings and practices of Sustainability. In the best case, the practices and meanings of Sustainability can constitute an element of reflexivity into the economy, science, politics, and everyday lifestyles. This reflexivity requires constant care including critical reflexivity. Moreover, it is a process rather than a state that could be achieved, and caring for Sustainability will not become obsolete.

Notes

1 See http://esango.un.org/civilsociety/login.do [accessed 17 July 2015].
2 Caring involves the element of active taking care as well as the element of value that is attached to the issue that is cared for. For another example, see Felt et al. (2013) who advocate caring for the relationship between science and society.

BIBLIOGRAPHY

Agrawala, S. (1998). Structural and process history of the Intergovernmental Panel on Climate Change. *Climatic Change*, 39(4), 621–642.

Alcott, B. (2005). Jevons' paradox. *Ecological Economics*, 54(1), 9–21.

Alcott, B. (2008). The sufficiency strategy: Would rich-world frugality lower environmental impact? *Ecological Economics*, 64(4), 770–786.

Alexander, S. (2014). *Degrowth and the Carbon Budget: Powerdown Strategies for Climate Stability*. Simplicity Institute.

Allen, P. & Kovach, M. (2000). The capitalist composition of organic: The potential of markets in fulfilling the promise of organic agriculture. *Agriculture and Human Values*, 17(3), 221–232.

Altieri, M. A. & Nicholis, C. I. (2005). *Agroecology and the Search for a Truly Sustainable Agriculture*. United Nations Environmental Programme, Environmental Training Network for Latin America and the Caribbean.

Aschemann, J., Hamm, U., Naspetti, S. & Zanoli, R. (2007). The organic market. In W. Lockeretz (ed.), *Organic Farming: An International History* (pp. 123–151). Wallington, Oxon: CAB International.

Atlason, R. & Unnthorsson, R. (2014). Ideal EROI (energy return on investment) deepens the understanding of energy systems. *Energy*, 67, 241–245.

Balfour, E. B. (1976). *The Living Soil and the Haughley Experiment*. New York: Universe Books.

Barbier, E. B. (1987). The concept of sustainable economic development. *Environmental Conservation*, 14(02), 101–110.

Barbier, E. B. (2009). *Rethinking the Economic Recovery: A Global Green New Deal*. UNEP.

Barney, G. O. (ed.) (1980). *The Global 2000 Report to the President of the U.S.: Entering the 21st Century*. New York: Pergamon Press.

Barnola, J. M., Raynaud, D., Korotkevich, Y. S. & Lorius, C. (1987). Vostok ice core provides 160,000-year record of atmospheric CO_2. *Nature*, 329 (6138), 408–414.

Barry, J. (2006). Resistance is fertile: From environmental to sustainability citizenship. In A. Dobson & D. Bell (eds), *Environmental Citizenship* (pp. 21–48). Cambridge, MA: MIT Press.

Barry, J. & Proops, J. (1999). Seeking sustainability discourses with Q methodology. *Ecological Economics*, 28(3), 337–345.

Barton, R. (1987). John Tyndall, Pantheist: A rereading of the Belfast Address. *Osiris*, 3, 111–134.

Beck, S. (2011). Moving beyond the linear model of expertise? IPCC and the test of adaptation. *Regional Environmental Change*, 11(2), 297–306.

Bengtsson, L. (2006). Geo-engineering to confine climate change: Is it at all feasible? *Climatic Change*, 77(3–4), 229–234.

Beretta, G. P. (2007). World energy consumption and resources: an outlook for the rest of the century. *International Journal of Environmental Technology and Management*, 7(1–2), 99–112.

Biello, D. (2010). Negating 'Climategate': Copenhagen talks and climate science survive stolen e-mail controversy. *Scientific American*, 302(2), 16.

Biermann, F. (2014). *Earth System Governance: World Politics in the Anthropocene*. Cambridge, MA: The MIT Press.

Blackstock, J. J., Battisti, D. S., Caldeira, K., Eardley, D. M., Katz, J. I., Keith, D. W., Patrinos, A. N., Schrag, D. P., Socolow, R. H. & Koonin, S. E. (2009). Climate engineering responses to climate emergencies (Novim). Retrieved 20 December 2015 from http://arxiv.org/pdf/0907.5140.

Blondell-Mégrelis, M. (2007). Liebig or how to popularize chemistry. *HYLE–International Journal for Philosophy of Chemistry*, 13(1), 43–54.

Borel-Saladin, J. M. & Turok, I. N. (2013). The Green Economy: Incremental change or transformation? *Environmental Policy and Governance*, 23(4), 209–220.

Bosello, F., Roson, R. & Tol, R. S. J. (2007). Economy-wide estimates of the implications of climate change: Sea level rise. *Environmental and Resource Economics*, 37(3), 549–571.

Boulding, K. E. (1966). The economics of the coming spaceship earth. In H. Jarret (ed.), *Environmental Quality in a Growing Economy* (pp. 3–14). Baltimore: Johns Hopkins University Press.

Bowker, G. C. & Leigh Star, S. (2000). *Sorting Things Out. Classification and its Consequences*. Cambridge, MA: MIT Press.

Brand, K.-W. (2011). Umweltsoziologie und der praxistheoretische Zugang. In M. Gross (ed.), *Handbuch Umweltsoziologie* (pp. 173–198). VS Verlag für Sozialwissenschaften.

Brown, V. A., Harris, J. A. & Russell, J. Y. (2010). *Tackling Wicked Problems through the Transdisciplinary Imagination*. Earthscan.

Buttel, F. H. (2000). Ecological modernization as social theory. *Geoforum*, 31(1), 57–65.

Callon, M. (1986). Some elements of a sociology of translation: Domestication of the scallops and the fishermen of St. Brieuc Bay. In J. Law (ed.), *Power, Action and Belief. A new Sociology of Knowledge?* (pp. 196–233). London: Routledge & Kegan Paul.

Carbon Tracker Initiative. (2012). *Unburnable Carbon – Are the World's Financial Markets Carrying a Carbon Bubble?* London: Carbon Tracker Initiative. Retrieved 20 December 2015 from http://www.carbontracker.org/wp-con tent/uploads/2014/09/Unburnable-Carbon-Full-rev2-1.pdf.

Carson, R. (1962). *Silent Spring*. Boston: Houghton Mifflin Harcourt.

Charney, J., Arakawa, A., Baker, J., Bolin, B., Dickinson, R. E., Goody, R. M., Leith, C. E., Stommel, H. W. & Wunsch, C. I. (1979). *Carbon Dioxide and Climate: A Scientific Assessment*. Washington, DC: National Academy of Sciences.

Chen, R. S., Boulding, E. & Schneider, S. H. (1983). *Social Science Research and Climate Change: An Interdisciplinary Appraisal*. Springer Science & Business Media.

Cook, J., Nuccitelli, D., Green, S. A., Richardson, M., Winkler, B., Painting, R., Way, R., Jacobs, P. & Skuce, A. (2013). Quantifying the consensus on anthropogenic global warming in the scientific literature. *Environmental Research Letters, 8*(2), 024024.

Daly, H. E. (1977). *Steady-state Economics: The Economics of Biophysical Equilibrium and Moral Growth*. San Francisco: W.H. Freeman.

'Declaration of Nyéléni' (2007). Retrieved 6 October 2015, from http://nye leni.org/spip.php?article290.

Deffeyes, K. S. (2008). *Hubbert's Peak: The Impending World Oil Shortage*. Princeton, NJ: Princeton University Press.

Demaria, F., Schneider, F., Sekulova, F. & Martinez-Alier, J. (2013). What is degrowth? From an activist slogan to a social movement. *Environmental Values*, 22(2), 191–215.

Depledge, J. (2000). *Tracing the Origins of the Kyoto Protocol: An Article-by-article Textual History: Technical Paper*. United Nations, UNFCCC Framework Convention on Climate Change.

Dinçer, İ. & Zamfirescu, C. (2012). *Sustainable Energy Systems and Applications*. New York: Springer US.

Dobson, A. (2007). Environmental citizenship: Towards sustainable development. *Sustainable Development*, 15(5), 276–285.

Dresner, S. (2012). *The Principles of Sustainability* (2nd edn). London: Routledge.

Du Pisani, J. A. (2006). Sustainable development – historical roots of the concept. *Environmental Sciences*, 3(2), 83–96.

Ekins, P. & Barker, T. (2001). Carbon taxes and carbon emissions trading. *Journal of Economic Surveys*, 15(3), 325–376.

Elkington, J. (1997). *Cannibals with Forks: The Triple Bottom Line of 21st Century Business*. Oxford: Capstone.

Ely, A., Smith, A., Stirling, A., Leach, M. & Scoones, I. (2013). Innovation politics post-Rio+20: Hybrid pathways to sustainability? *Environment and Planning C: Government and Policy*, 31(6), 1063–1081.

Emerich, M. M. (2011). *The Gospel of Sustainability: Media, Market, and LOHAS*. University of Illinois Press.

European Commission. (2014a). *Action Plan for the Future of Organic Production in the European Union*. Brussels.

European Commission. (2014b). *Energy Efficiency and its Contribution to Energy Security and the 2030 Framework for Climate and Energy Policy* (No. COM (2014) 520 final). Brussels.

Fairtrade International. (2010). Fairtrade's contribution to a more sustainable world (Position Paper). Retrieved 20 December 2015 from http://www.fairtrade.net/fileadmin/user_upload/content/2009/resources/2010-12-31_flo-sustainability-position-paper.pdf.

Federal Ministry of Economics and Technology. (2012). Germany's new energy policy (Special Brochure. Spotlight on Economic Policy). Retrieved 20 December 2015 from http://www.bmwi.de/English/Redaktion/Pdf/germanys-new-energy-policy.

Felt, U., Barben, D., Irwin, A., Joly, P.-B., Rip, A., Stirling, A., Stöckelová, T. (2013). *Science in Society: Caring for Our Futures in Turbulent Times*. European Science Foundation. Retrieved 20 December 2015 from http://www.esf.org/fileadmin/Public_documents/Publications/spb50_ScienceInSociety.pdf.

Felt, U. & Fochler, M. (2010). Machineries for making publics: Inscribing and de-scribing publics in public engagement. *Minerva*, 48(3), 219–238.

Felt, U., Igelsböck, J., Schikowitz, A. & Voelker, T. (2012). Challenging participation in sustainability research. *International Journal of Deliberative Mechanisms in Science* (1), 4–34.

Flear, M. L. & Pfister, T. (2015). Contingent participation: Imaginaries of sustainable technoscientific innovation in the European Union. In M. D. Pickersgill & E. Cloatre (eds), *Knowledge, Technology and Law: Interrogating the Nexus* (pp. 33–49). London: Routledge.

Fleck, L. (1980). *Entstehung und Entwicklung einer wissenschaftlichen Tatsache: Einfuehrung in die Lehre vom Denkstil und Denkkollektiv*. Frankfurt am Main: Suhrkamp.

Fleming, J. R. (1999). Joseph Fourier, the 'greenhouse effect', and the quest for a universal theory of terrestrial temperatures. *Endeavour*, 23(2), 72–75.

Fleming, J. R. (2005). *Historical Perspectives on Climate Change*. Oxford University Press.

Forsyth, T. (2003). *Critical Political Ecology: the Politics of Environmental Science*. London: Routledge.

Friedman, T. L. (2007). The power of green. *The New York Times*, 15 April. Retrieved 20 December 2015 from http://www.nytimes.com/2007/04/15/magazine/15green.t.html.

Gallie, W. B. (1956). Essentially contested concepts. *Proceedings of the Aristotelian Society*, 56 (ns), 167–198.

Geier, B. (2007). IFOAM and the history of the international organic movement. In W. Lockeretz (ed.), *Organic Farming: An International History* (pp. 175–186). Wallington, Oxon: CAB International.

Groß, M. & Mautz, R. (2015). *Renewable Energies*. London: Routledge.

Grubb, M. (2003). The economics of the Kyoto Protocol. *World Economics*, 4(3), 143–190.

Hall, C. A. S. & Day Jr, J. W. (2009). Revisiting the limits to growth after peak oil. *American Scientist*, 97(3), 230–237.

Hall, C. A. S., Lambert, J. G. & Balogh, S. B. (2014). EROI of different fuels and the implications for society. *Energy Policy*, 64, 141–152.

Hallegatte, S., Heal, G., Fay, M. & Treguer, D. (2012). From growth to green growth – a framework (Working Paper No. 17841). National Bureau of Economic Research. Retrieved 20 December 2015 from http://www.nber.org/papers/w17841.

Hanna, A. & Lacy, P. (2011). *Towards a New Era of Sustainability in the Energy Industry*. UN Global Compact-Accenture.

Hansen, J. W. (1996). Is agricultural sustainability a useful concept? *Agricultural Systems*, 50(2), 117–143.

Haraway, D. (1988). Situated knowledges: The science question in feminism and the privilege of partial perspective. *Feminist Studies*, 14(3), 575–599.

Hardin, G. (1968). The Tragedy of the Commons. *Science* 162(3859), 1243–1248.

Harwood, R. R. (1990). A history of sustainable agriculture. In C. A. Edwards, R. Lal, P. Madden, R. H. Miller & G. House (eds), *Sustainable Agricultural Systems* (pp. 3–19). Florida: St. Lucie Press.

Hazell, P. B. R. (2003). The green revolution. In J. Mokyr (ed.), *The Oxford Encyclopedia of Economic History* (pp. 478–480). Oxford, UK: Oxford University Press.

Hecht, G. (2009). *The Radiance of France: Nuclear Power and National Identity after World War II*. Cambridge, MA: MIT Press.

Heinberg, R. & Bomford, M. (2009). *The Food and Farming Transition: Toward a Post Carbon Food System*. Sebastopol, USA: Post Carbon Institute.

HM Government. (2013a, 26 March). Long-term nuclear energy strategy. Department for Business, Innovation & Skills; Department of Energy & Climate Change. Retrieved 20 December 2015 from https://www.gov.uk/

government/uploads/system/uploads/attachment_data/file/168047/bis-13-630-long-term-nuclear-energy-strategy.pdf.

HM Government. (2013b, 26 March). Nuclear industrial strategy: the UK's nuclear future. Department for Business, Innovation & Skills; Department of Energy & Climate Change. Retrieved 20 December 2015 from https://www.gov.uk/government/uploads/system/uploads/attachment_data/file/168048/bis-13-627-nuclear-industrial-strategy-the-uks-nuclear-future.pdf.

Hockerts, K. (2005). *The Fair Trade Story*. Retrieved 20 December 2015 from http://www.fairtrade.at/fileadmin/user_upload/PDFs/Fuer_Studierende/oikos_winner2_2005.pdf.

Holling, C. S. (1973). Resilience and stability of ecological systems. *Annual Review of Ecology and Systematics*, 4, 1–23.

Hoppmann, J., Huenteler, J. & Girod, B. (2014). Compulsive policy-making – The evolution of the German feed-in tariff system for solar photovoltaic power. *Research Policy*, 43(8), 1422–1441.

Houghton, J., Ding, Y., Griggs, S. J., Noguer, M., van der Linden, P. J., Dai, X., Maskell, A. & Johnson, C. A. (eds). (2001). *Climate Change 2001: The Scientific Basis: Contribution of Working Group I to the Third Assessment Report of the Intergovernmental Panel on Climate Change*. Cambridge: Cambridge University Press.

Houghton, J. T., Meira Filho, L. G., Callander, B. A., Harris, N., Kattenberg, A. & Maskell, K. (1996). *Climate Change 1995: The Science of Climate Change: Contribution of Working Group I to the Second Assessment Report of the Intergovernmental Panel on Climate Change*. Cambridge: Cambridge University Press.

Howard, A. S. (1943). *An Agricultural Testament*. New York: Oxford University Press.

Hulme, M. (2014). *Can Science Fix Climate Change? A Case against Climate Engineering*. Cambridge and Malden, MA: Polity Press.

Hulme, M. & Mahony, M. (2010). Climate change: What do we know about the IPCC? *Progress in Physical Geography*, 34(5), 705–718.

Humphery, K. (2015). Sustainable consumption. In Cook, D. T. & Ryan, J. M. (eds) *The Wiley Blackwell Encyclopedia of Consumption and Consumer Studies*. Chichester and New York: Wiley-Blackwell.

IAASTD. (2009). *Agriculture at a Crossroads. International Assessment of Agricultural Knowledge, Science and Technology for Development. Global Report*. Washington, DC: Island Press.

IFOAM. (2012). *Organic without Boundaries. IFOAM celebrating 40 Years*. Bonn: IFOAM.

Intergovernmental Panel on Climate Change. (2007). *Climate Change 2007: Synthesis Report* (IPCC Synthesis Report). Geneva. Retrieved 20 December 2015 from http://www.ipcc.ch/pdf/assessment-report/ar4/syr/ar4_syr.pdf.

Intergovernmental Panel on Climate Change. (2015). *Climate Change 2014: Synthesis Report* (IPCC Synthesis Report). Geneva. Retrieved 20 December

2015 from http://www.ipcc.ch/pdf/assessment-report/ar5/syr/SYR_AR5_FINAL_full.pdf.

International Energy Agency. (2010a). *Energy Poverty: How to Make Modern Energy Access Universal?*Paris, France.

International Energy Agency. (2010b). *World Energy Outlook 2010.* Paris, France.

International Energy Agency. (2014). *Key World Energy Statistics.* Paris, France.

IPCC. (2011). *Special Report on Renewable Energy Sources and Climate Change Mitigation. Prepared by Working Group III of the Intergovernmental Panel on Climate Change.* Eds O. Edenhofer, R. Pichs-Madruga, Y. Sokona,, K. Seyboth, P. Matschoss, S. Kadner, T. Zwickel, P. Eickemeier, G. Hansen, & C. von Stechow. United Kingdom and New York, NY, USA: Cambridge University Press. Retrieved 20 December 2015 from http://srren.ipcc-wg3.de/report.

Irwin, A. (2006). The politics of talk: Coming to terms with the 'new' scientific governance. *Social Studies of Science*, 36(2), 299–320.

IUCN. (1980). *World Conservation Strategy. Living Resource Conservation for Sustainable Development.* International Union for Conservation of Nature.

Jackson, R. B., Vengosh, A., Carey, J. W., Davies, R. J., Darrah, T. H., O'Sullivan, F. & Pétron, G. (2014). The environmental costs and benefits of fracking. *Annual Review of Environment and Resources*, 39(1), 327–362.

Jackson, T. (2005). Live better by consuming less: Is there a 'double dividend' in sustainable consumption. *Journal of Industrial Ecology*, 9(1–2), 19–36.

Jackson, T. (2009). *Prosperity without Growth? The Transition to a Sustainable Economy.* Sustainable Development Commission.

Jackson, W. (1980). *New Roots for Agriculture.* University of Nebraska Press.

Jasanoff, S. (2001). Image and imagination: The formation of global environmental consciousness. In P. Edwards & C. Miller (eds), *Changing the Atmosphere: Expert Knowledge and Environmental Governance* (pp. 309–337). Cambridge, MA: MIT Press.

Jasanoff, S. (2004). Ordering knowledge, ordering society. In S. Jasanoff (ed.), *States of Knowledge: The Co-Production of Science and Social Order* (pp. 13–45). London: Routledge.

Jasanoff, S. (2012). *Science and Public Reason.* London; New York: Routledge.

Jasanoff, S. & Kim, S.-H. (2009). Containing the atom: Sociotechnical imaginaries and nuclear power in the United States and South Korea. *Minerva*, 47(2), 119–146.

Jevons, W. S. (1865). *The Coal Question: An Enquiry Concerning the Progress of the Nation, and the Probable Exhaustion of Our Coal-mines.* Macmillan.

Keeling, R. F. (2008). Recording Earth's vital signs. *Science*, 319, 1771–1772.

King, D. A. (2004). Climate change science: Adapt, mitigate, or ignore? *Science*, 303(5655), 176–177.

Kirschenmann, F. (2014). A brief history of sustainable agriculture. *The Networker*, 9(2). Retrieved 20 December 2015 from http://www.sehn.org/Volume_9-2.html.

Knorr Cetina, K. (1999). *Epistemic Cultures. How the Sciences Make Knowledge*. Cambridge, MA: Harvard University Press.

Knorr Cetina, K. (2001). Objectual practice. In R. Theodore Schatzki, K. Knorr-Cetina & E. Von Savigny (eds), *The Practice Turn*. London: Routledge.

Koc, M. (2010). Sustainability: A tool for food system reform. In A. Blay-Palmer (ed.), *Imagining Sustainable Food Systems: Theory and Practice* (pp. 37–45). Surrey, UK: Ashgate Publishing, Ltd.

Kuhn, T. (1962). *The Structure of Scientific Revolutions*. Chicago: University of Chicago Press.

Lang, T. & Barling, D. (2013). Nutrition and sustainability: An emerging food policy discourse. *Proceedings of the Nutrition Society*, 72(01), 1–12.

Latouche, S. (2010). Degrowth. *Journal of Cleaner Production*, 18(6), 519–522.

Latour, B. (1987). *Science in Action. How to Follow Scientists and Engineers through Society*. Cambridge, MA: Harvard University Press.

Latour, B. (2005). *Reassembling the Social: An Introduction to Actor-Network-Theory*. Oxford: Oxford University Press.

Leiserowitz, A. A., Maibach, E. W., Roser-Renouf, C., Smith, N. & Dawson, E. (2013). Climategate, public opinion, and the loss of trust. *American Behavioral Scientist*, 57(6), 818–837.

Lockeretz, W. (ed.). (2007a). *Organic Farming: An International History*. Wallington, Oxon: CAB International.

Lockeretz, W. (2007b). What explains the raise of organic farming. In W. Lockeretz (ed.), *Organic Farming: An International History* (pp. 1–8). Wallington, Oxon: CAB International.

Luber, G. & McGeehin, M. (2008). Climate change and extreme heat events. *American Journal of Preventive Medicine*, 35(5), 429–435.

Lynch, M. (1993). *Scientific Practice and Ordinary Action: Ethnomethodology and Social Studies of Science*. Cambridge: Cambridge University Press.

Mackenzie, F. T. (1998). *Our Changing Planet: An Introduction to Earth System Science and Global Environmental Change*. Prentice Hall.

Mann, M. E. (2013). *The Hockey Stick and the Climate Wars: Dispatches from the Front Lines*. Columbia University Press.

Mann, M. E., Bradley, R. S. & Hughes, M. K. (1999). Northern hemisphere temperatures during the past millennium: Inferences, uncertainties, and limitations. *Geophysical Research Letters*, 26(6), 759–762.

Mann, M. E., Zhang, Z., Rutherford, S., Bradley, R. S., Hughes, M. K., Shindell, D., Ammann, C., Faluvegi, G. & Ni, F. (2009). Global signatures and dynamical origins of the Little Ice Age and Medieval Climate Anomaly. *Science*, 326(5957), 1256–1260.

McMichael, A. J., Woodruff, R. E. & Hales, S. (2006). Climate change and human health: present and future risks. *The Lancet*, 367(9513), 859–869.

Meadows, D. H., Meadows, D. L., Randers, J. & Behrens, W. W. (1972). *The Limits to Growth. A Report for the Club of Rome's Project on the Predicament of Mankind*. New York: Universe Books.

Meinshausen, M., Meinshausen, N., Hare, W., Raper, S. C. B., Frieler, K., Knutti, R., Frame, D. J. & Allen, M. R. (2009). Greenhouse-gas emission targets for limiting global warming to 2°C. *Nature*, 458(7242), 1158–1162.

Millennium Ecosystem Assessment. (2005). *Ecosystems and Human Well-being: Synthesis*. Washington, DC: Island Press. Retrieved 20 December 2015 from http://www.millenniumassessment.org/documents/document.356.aspx.pdf.

Miller, C. A. (2005). New civic epistemologies of quantification: Making sense of local and global indicators of sustainability. *Science, Technology & Human Values*, 30(3), 403–432.

Monsanto. (2015). Our Commitments. Retrieved 16 June 2015, from http://www.monsanto.com/whoweare/pages/our-commitments.aspx.

Morisse-Schilbach, M. & Halfmann, J. (2012). Einführung: Die Herausforderungen von Wissen(schaft) und Politik jenseits des Staates durch Global Commons. In M. Morisse-Schilbach & J. Halfmann (eds), *Wissen, Wissenschaft und Global Commons. Konturen eines interdisziplinären Forschungsfeldes* (pp. 15–58). Baden-Baden: Nomos.

Murphy, D. J. & Hall, C. A. S. (2011). Energy return on investment, peak oil, and the end of economic growth. *Annals of the New York Academy of Sciences*, 1219(1), 52–72.

National Research Council. (1979). *Carbon Dioxide and Climate: A Scientific Assessment*. Washington, DC: The National Academies Press.

Nelkin, D. (1979). *Controversy. Politics of Technical Decisions*. London: Sage Publications.

Nelson, V. C. (2011). *Introduction to Renewable Energy*. CRC Press.

Nemeth, D. F. (2013). Thoughts on Katrina vs. Sandy: Donald F. Nemeth. *Ecopsychology*, 5(S1), S–4.

Nerlich, B. (2010). 'Climategate': Paradoxical metaphors and political paralysis. *Environmental Values*, 19(4), 419–442.

New Scientist. (2007). Climate myths: The cooling after 1940 shows CO_2 does not cause warming, 16 May. Retrieved 20 December 2015 from https://www.newscientist.com/article/dn11639-climate-myths-the-cooling-after-1940-shows-co2-does-not-cause-warming/.

Nicholls, R. J., Marinova, N., Lowe, J. A., Brown, S., Vellinga, P., Gusmão, D. de, Hinkel, J. & Tol, R. S. J. (2011). Sea-level rise and its possible impacts given a 'beyond 4°C world' in the twenty-first century. *Philosophical Transactions of the Royal Society of London A: Mathematical, Physical and Engineering Sciences*, 369(1934), 161–181.

Nordhaus, W. D. (1975). Can we control carbon dioxide? IIASA Working Paper WP-75-063. International Institute for Applied Systems Analysis: Laxenburg.

Norman, W. & MacDonald, C. (2004). Getting to the bottom of 'triple bottom line'. *Business Ethics Quarterly*, 14(02), 243–262.

Northbourne, L. W. J. (1940). *Look to the Land*. London: Dent.

OECD. (2011). *OECD Green Growth Studies Fostering Innovation for Green Growth*. OECD Publishing.

Oldeman, L. R. (1992). The global extent of soil degradation. In *ISRIC Bi-Annual Report 1991–1992* (pp. 19–36). Wageningen, The Netherlands.

O'Riordan, T. (2004). Environmental science, sustainability, and politics. *Transactions of the Institute of British Geographers*, New Series (29), 234–247.

Paarlberg, D. & Paarlberg, P. (2000). *The Agricultural Revolution of the 20th Century* (1st edn). Ames: Iowa State University Press.

Pallett, H. & Chilvers, J. (2013). A decade of learning about publics, participation, and climate change: Institutionalising reflexivity? *Environment and Planning A*, 45(5), 1162–1183.

Paull, J. (2010). From France to the world: The International Federation of Organic Agriculture Movements (IFOAM). *Journal of Social Research and Policy*, 1(2), 93–102.

Paull, J. (2014). Lord Northbourne, the man who invented organic farming, a biography. *Journal of Organic Systems*, 9(1), 31–53.

Petit, J. R., Jouzel, J., Raynaud, D., Barkov, N. I., Barnola, J.-M., Basile, I., Bender, M., Chappellaz, J., Davis, M., Delaygue, G., Delmotte, M., Kotlyakov, V. M., Legrand, M., Lipenkov, V. Y., Lorius, C., Pépin, L., Ritz, C., Saltman, E., & Stievenard, M. (1999). Climate and atmospheric history of the past 420,000 years from the Vostok ice core, Antarctica. *Nature*, 399 (6735), 429–436.

Pickering, A. (1995). *The Mangle of Practice. Time, Agency, and Science*. Chicago: University of Chicago Press.

Pielke, R. A. (2007). *The Honest Broker: Making Sense of Science in Policy and Politics*. Cambridge and New York: Cambridge University Press.

Pielke, R., Prins, G., Rayner, S. & Sarewitz, D. (2007). Climate change 2007: Lifting the taboo on adaptation. *Nature*, 445(7128), 597–598.

Pilcher, J. (2006). *Food in World History*. New York and Abingdon: Routledge.

Porter, T. M. (1995). *Trust in Numbers: the Pursuit of Objectivity in Science and public Life*. Princeton, NJ: Princeton University Press.

Princen, T. (2003). Principles for sustainability: From cooperation and efficiency to sufficiency. *Global Environmental Politics*, 3(1), 33–50.

Princen, T. (2005). *The Logic of Sufficiency*. Cambridge, MA: The MIT Press.

Radkau, J. (2011). *Die Ära der Ökologie: eine Weltgeschichte*. Bonn: Bundeszentrale für Politische Bildung.

Rahmstorf, S. (2007). A semi-empirical approach to projecting future sea-level rise. *Science*, 315(5810), 368–370.

Ravetz, J. R. (2011). 'Climategate' and the maturing of post-normal science. *Futures*, 43(2), 149–157.

Raynolds, L. T. (2000). Re-embedding global agriculture: The international organic and fair trade movements. *Agriculture and Human Values*, 17(3), 297–309.

Reckwitz, A. (2002). Toward a theory of social practices. A development in culturalist theorizing. *European Journal of Social Theory*, 5(2), 243–263.

Redfern, A. & Snedker, P. (2002). Creating market opportunities for small enterprises: Experiences of the Fair Trade Movement (SEED Working Paper No. 30). International Labour Organization.

Rip, A., Misa, T. J. & Schot, J. (1995). *Managing Technology in Society: The Approach of Constructive Technology Assessment*. London: Pinter Publishers.

Rittel, H. W. J. & Webber, M. M. (1973). Dilemmas in a general theory of planning. *Policy Sciences*, 4, 155–169.

Rodhe, H., Charlson, R. & Crawford, E. (1997). Svante Arrhenius and the Greenhouse Effect. *Ambio*, 26(1), 2–5.

Samuel, G. (2013). *25 Years Of EUROSOLAR 1988–2013. From Seizing the Initiative to Being A Pioneer. From Vision to Practice*. Ed. I. Scheer-Pontenagel. EUROSOLAR: Bonn.

Sarewitz, D. (2004). How science makes environmental controversies worse. *Environmental Science and Policy*, 7, 385–403.

Schmidt, G. (2010). Carbon dioxide and the climate. *American Scientist*, 98(1), 58.

Scoones, I., Leach, M. & Newell, P. (eds) (2015). *The Politics of Green Transformations*. London and New York: Routledge.

Shove, E. (2003). Converging conventions of comfort, cleanliness and convenience. *Journal of Consumer Policy*, 26(4), 395–418.

Shove, E., Pantzar, M. & Watson, M. (2012). *The Dynamics of Social Practice Everyday Life and How it Changes*. Los Angeles: SAGE.

Shove, E., Watson, M. & Spurling, N. (2015). Conceptualizing connections: Energy demand, infrastructures and social practices. *European Journal of Social Theory*, 18(3), 274–287.

Singer, S. F. (2000). *Climate Policy – From Rio to Kyoto* (Vol. 102). Stanford: Hoover Press.

Smil, V. (2004a). *Enriching the Earth: Fritz Haber, Carl Bosch, and the Transformation of World Food Production*. Cambridge, MA: The MIT Press.

Smil, V. (2004b). World history and energy. *Encyclopedia of Energy*, 6, 549–561.

Smil, V. (2011). Nitrogen cycle and world food production. *World Agriculture*, 2, 9–13.

Smith, A. (2012). Civil society in sustainable energy transitions. In G. Verbong & D. Loorbach (eds), *Governing the Energy Transition: Reality, Illusion, or Necessity*. New York: Routledge.

Smith, A. & Stirling, A. (2010). The politics of social-ecological resilience and sustainable socio-technical transitions. *Ecology and Society*, 15(1), 11.

Sørensen, B. (1991). A history of renewable energy technology. *Energy Policy*, 19(1), 8–12.

Stern, N. (2007). *The Economics of Climate Change: The Stern Review*. Cambridge, UK and New York: Cambridge University Press.

Stirling, A. (2008). 'Opening up' and 'closing down': Power, participation, and pluralism in the social appraisal of technology. *Science, Technology & Human Values*, 33(2), 262–294.

Stirling, A. (2009). Participation, precaution, and reflexive governance for sustainable development. In W. N. Adger & A. Jordan (eds), *Governing Sustainability* (pp. 193–225). Cambridge: Cambridge University Press.

Tauger, M. B. (2011). *Agriculture in World History*. Abingdon and New York: Routledge.

Thee-Brenan, M. (2014). Americans are outliers in views on climate change. *The New York Times*, 6 May. Retrieved 20 December 2015 from http://www.nytimes.com/2014/05/07/upshot/americans-are-outliers-in-views-on-climate-change.html.

Toke, D. (2011). Ecological modernisation, social movements and renewable energy. *Environmental Politics*, 20(1), 60–77.

Tollefson, J. (2015). Climate-change 'hiatus' disappears with new data. *Nature*. Retrieved 6 October 2015, from http://www.nature.com/news/climate-change-hiatus-disappears-with-new-data-1.17700.

UNEP. (2008). Global Green New Deal – UNEP Green Economy Initiative – Press Releases October 2008. Retrieved 7 July 2015, from http://www.unep.org/Documents.Multilingual/Default.asp?DocumentID=548&ArticleID=5955&l=en.

UNEP. (2009). *Global Green New Deal*. United Nations Environment Programme.

Union of Concerned Scientists. (2015). A short history of energy. Retrieved 23 July 2015, from http://www.ucsusa.org/clean_energy/our-energy-choices/a-short-history-of-energy.html.

United Nations. (1992a). *Rio Declaration on Environment and Development*. Retrieved 20 December 2015 from http://www.un.org/documents/ga/conf151/aconf15126-1annex1.htm.

United Nations (ed.) (1992b). *United Nations Framework Convention on Climate Change*. United Nations Publications: Geneva, Switzerland.

US–Canada Power System Outage Task Force. (2004). *Final Report on the August 14, 2003 Blackout in the United States and Canada: Causes and Recommendations*. Retrieved 20 December 2015 from http://energy.gov/sites/prod/files/oeprod/DocumentsandMedia/BlackoutFinal-Web.pdf.

van der Vleuten, E. & Lagendijk, V. (2010). Transnational infrastructure vulnerability: The historical shaping of the 2006 European 'Blackout'. *Energy Policy*, 38(4), 2042–2052.

Vanek, F. & Albright, L. (2008). *Energy Systems Engineering: Evaluation and Implementation*. New York: McGraw Hill Professional.

Vermeulen, S. J., Campbell, B. M. & Ingram, J. S. I. (2012). Climate change and food systems. *Annual Review of Environment and Resources*, 37(1), 195–222.

Vogt, G. (2007). The origins of organic farming. In W. Lockeretz (ed.), *Organic Farming: An International History* (pp. 9–29). Wallington, Oxon: CAB International.

Voss, J.-P., Bauknecht, D. & Kemp, R. (2006). *Reflexive Governance for Sustainable Development*. Cheltenham: Edward Elgar.

Voss, J.-P. & Simons, A. (2014). Instrument constituencies and the supply side of policy innovation: The social life of emissions trading. *Environmental Politics*, 23(5), 735–754.

Walker, B. & Salt, D. (2012). *Resilience Thinking: Sustaining Ecosystems and People in a Changing World*. Washington: Island Press.

WCED. (1987). *Report of the World Commission on Environment and Development: Our Common Future* (UN Document). Retrieved 20 December 2015 from http://www.un-documents.net/our-common-future.pdf.

Weart, S. (2013). Rise of interdisciplinary research on climate. *Proceedings of the National Academy of Sciences*, 110(Supplement 1), 3657–3664.

Welzer, H. (2012). *Climate Wars: What People Will Be Killed For in the 21st Century* (1st edn). Cambridge, UK and Malden, MA: John Wiley & Sons.

Wheeler, T. & Braun, J. von. (2013). Climate change impacts on global food security. *Science*, 341(6145), 508–513.

Whitmarsh, L. (2011). Scepticism and uncertainty about climate change: Dimensions, determinants and change over time. *Global Environmental Change*, 21(2), 690–700.

Wittgenstein, L. (1953). *Philosophical Investigations*. Oxford: Wiley-Blackwell.

Woods, J., Williams, A., Hughes, J. K., Black, M. & Murphy, R. (2010). Energy and the food system. *Philosophical Transactions of the Royal Society of London B: Biological Sciences*, 365(1554), 2991–3006.

World Bank. (2012). *Inclusive Green Growth: The Pathway to Sustainable Development*. The World Bank. Retrieved 20 December 2015 from http://elibrary.worldbank.org/doi/book/10.1596/978-0-8213-9551-6.

Wynne, B. (2010). Strange weather, again: Climate science as political art. *Theory, Culture & Society*, 27(2–3), 289–305.

Yaritani, H. & Matsushima, J. (2014). Analysis of the energy balance of shale gas development. *Energies*, 7(4), 2207–2227.

Yin, J. H. & Battisti, D. S. (2001). The importance of tropical sea surface temperature patterns in simulations of last glacial maximum climate. *Journal of Climate*, 14(4), 565–581.

INDEX

Absorption capacity 15
Access 9, 32, 48–9, 54, 57
Accounting 20, 59, 82, 89
Agenda 21 19

Balfour, Lady Eve 68
Ban Ki-Moon 57
Biodynamic agriculture/biodynamic
    farming 67, 69, 77n4
Brundtland, Gro Harlem 7, 17
Brundtland Report, see *World
    Commission on Environment and
    Development*
BSE 68

Cap-and-trade 34–5
Carbon bubble 48
Care/caring, *for the environment* 13, 41,
    54, 67, 69
Care/caring, *for Sustainability* 89–93
Carlowitz, Hans Carl von 13
Carson, Rachel 14–15, 69
Climate change 48, 57–8, 63, 65, 79,
    83, 85–6, 88–90

Climate engineering 40, 42, 92
Club of Rome 15
Coal 13, 27, 44–5, 49, 51
Collapse 15
Columella, Lucius Junius Moderatus 13
Conferences of the Parties (COP) 34–5
Corporate social responsibility (CSR) 1

Daly, Herman 7
DDT 14
Declaration of Nyéléni 72
Degrowth 41–2, 56
Democracy 10–11, 83–4
Diversity (meanings and practices of
    Sustainability) 60, 76, 78, 81, 86
Du Pisani, Jacobus 13

Earth Summit, see *United Nations
    Conference for Environment and
    Development*
Ecological limits 15, 18, 40, 44
Electricity 45, 59
Elkington, John 19
Energy poverty 48

Energy system 43–9, 51–5, 57–8
Epistemic commons 82–4
Equilibrium 8, 15, 40
European Union/EU 23, 35, 69, 82, 88
EU 2020 strategy 23
Evelyn, John 13
Experiment/experimenting 82, 85–6, 88, 91–2

Fair trade 70–4, 76, 79–80
Fairtrade International (organisation) 71
Fleck, Ludwik 81
Food sovereignty 64, 66, 72–4
Forestry 13
Fossil 34, 44–53, 58, 64–5
Foucault, Michel 25n1
Fracking, see *shale gas*
Friedman, Thomas L. 22, 38
Friends of the Earth 15–16, 73, 86

Gas 27–8, 45, 50–1
Genealogy/genealogical 12, 20–1, 25n1
Genetically modified organisms (GMO) 76, 92
Green economy 22–25, 38, 42, 74, 79
Green growth 21, 23, 39, 42, 57, 74
Green new deal 22–3, 38
Greenwashing 60
Growth, economic 38, 40, 42, 56–9, 75, 79, 91–2
Growth, material 18–19, 42

Hardin, Garret 15
Haughley experiment 68
Health 47, 54–5, 62, 66, 74–6
Howard, Sir Albert 67–8
Hubbert, Marion King 49–50

International Assessment of Agricultural Knowledge, Science and Technology for Development (IAASTD) 66

Indicators 19–21, 81
Infrastructures 85–8, 90
Innovation 12, 23, 58–9
Institution 11, 14, 20–1, 29, 32, 67–9, 80–4, 87–8
Instruments 34, 73, 79–81, 88–90, 92
International Energy Agency (IEA) 48, 50
International Federation of Organic Agriculture Movements (IFOAM) 17, 68–9
International Panel on Climate Change (IPCC) 21, 29–31, 35, 37, 47, 53, 80, 83

Jevons, William 49, 56
Johannesburg 23

Knowledge 26–8, 30, 36, 38, 80–1, 87, 90
Kyoto Protocol 19, 21, 34–5, 39

La Via Campesina 72
Limits to Growth (report) 7, 15–16, 33, 41, 52, 57

Malthus, Thomas Robert 13
Massachusetts Institute of Technology (MIT) 15
Measuring 20, 37
Mill, John Stewart 13
Millennium Development Goals (MDGs) 23–4

Northbourne, Lord 77n4

Oil crisis 51
Oil, peak: *see* Peak oil
Organic farming 63, 68–9, 72–5, 77n4
Our common Future, see World Commission on Environment and Development
OXFAM 70

Peak oil 50
Practice theory 4

Radkau, Joachim 14
Rebound effect 13, 56
Redistribution 23, 25
Reflexivity 79, 82, 85, 90, 93
Renewable energy 44, 46–7, 52–9, 82, 88
Resilience 41–2
Rio Declaration on Environment and Development 19
Rio de Janeiro 19, 69

Shale gas 50
Spaceship Earth 15
Stationary state 13
Steady-state economy 7
Steiner, Rudolf 67–8, 77n4
Stern Review on the Economics of Climate Change 22, 38
Stirling, Andrew 8
Sufficiency 40–2, 92
Sustainable Development Goals (SDGs) 24

Technological voluntarism 39, 41–2, 79
Ten Thousand Villages 70

Transformation 20, 41, 46, 58, 79, 82, 85, 88–9
Triple bottom line 19, 59

United Nations Conference for Environment and Development (UNCED) 19
United Nations Environment Programme (UNEP) 17, 21–2, 29
United Nations Framework Convention on Climate Change (UNFCC) 19, 21, 34, 39
United Nations Sustainable Development Summit 24
U.S. Department of Agriculture (USDA) 68

Varro, Marcus Terentius 13
Via Campesina 72

Well-being 12, 20–1, 41, 47, 54, 62, 72, 78, 82–3
Wittgenstein, Ludwig 6, 81
World Commission on Environment and Development (WCED) 7, 17, 80

 Taylor & Francis eBooks

Helping you to choose the right eBooks for your Library

Add Routledge titles to your library's digital collection today. Taylor and Francis ebooks contains over 50,000 titles in the Humanities, Social Sciences, Behavioural Sciences, Built Environment and Law.

Choose from a range of subject packages or create your own!

Benefits for you

» Free MARC records
» COUNTER-compliant usage statistics
» Flexible purchase and pricing options
» All titles DRM-free.

REQUEST YOUR FREE INSTITUTIONAL TRIAL TODAY **Free Trials Available** We offer free trials to qualifying academic, corporate and government customers.

Benefits for your user

» Off-site, anytime access via Athens or referring URL
» Print or copy pages or chapters
» Full content search
» Bookmark, highlight and annotate text
» Access to thousands of pages of quality research at the click of a button.

eCollections – Choose from over 30 subject eCollections, including:

Archaeology	Language Learning
Architecture	Law
Asian Studies	Literature
Business & Management	Media & Communication
Classical Studies	Middle East Studies
Construction	Music
Creative & Media Arts	Philosophy
Criminology & Criminal Justice	Planning
Economics	Politics
Education	Psychology & Mental Health
Energy	Religion
Engineering	Security
English Language & Linguistics	Social Work
Environment & Sustainability	Sociology
Geography	Sport
Health Studies	Theatre & Performance
History	Tourism, Hospitality & Events

For more information, pricing enquiries or to order a free trial, please contact your local sales team:
www.tandfebooks.com/page/sales

 Routledge
Taylor & Francis Group

The home of
Routledge books

www.tandfebooks.com

Printed and bound by CPI Group (UK) Ltd, Croydon, CR0 4YY

22/10/2024

01777628-0002